Modelling and Simulation in Sport and Exercise

Modelling and simulation techniques are of central importance to conducting research in sport and exercise science, informing data collection and helping to analyze patterns of movement and physical performance. *Modelling and Simulation in Sport and Exercise* is the first book to offer an instructive reference for modelling and simulation methods for researchers and sport and exercise scientists.

Based around a series of research cases, describing core theories in applied, practical settings, the book draws on examples of modelling and simulation in ball games, biomechanical analysis, physiological testing and monitoring, predictive analysis and sports engineering and product design. Each research case presents a central problem, discusses different modelling approaches that could be used to deal with the issue, analysis of results and a reflection on the methodology and an exercise for students to put the techniques discussed into practice.

This is an important reference for any active researcher or upper-level student in sport and exercise science with an interest in mathematical modelling, computer science or simulation techniques.

Arnold Baca is head of the Section of Biomechanics, Kinesiology and Applied Computer Science in the Department of Sport Science at the University of Vienna, Austria.

Jürgen Perl is a Professor Emeritus in the Institute of Computer Science at the University of Mainz, Germany.

Routledge Research in Sport and Exercise Science

For more information about this series, please visit:
www.routledge.com/sport/series/RRSES

The *Routledge Research in Sport and Exercise Science* series is a showcase for cutting-edge research from across the sport and exercise sciences, including physiology, psychology, biomechanics, motor control, physical activity and health, and every core sub-discipline. Featuring the work of established and emerging scientists and practitioners from around the world, and covering the theoretical, investigative and applied dimensions of sport and exercise, this series is an important channel for new and groundbreaking research in the human movement sciences.

Available in this series:

Detecting Doping in Sport
Stephen Moston and Terry Engelberg

The Science of Climbing and Mountaineering
Edited by Ludovic Seifert, Peter Wolf and Andreas Schweizer

The Athlete Apperception Technique
Manual and Materials for Sport and Clinical Psychologists
Petah M. Gibbs, Mark B. Andersen and Daryl B. Marchant

Complex Sport Analytics
Felix Lebed

The Science of Figure Skating
Edited by Jason Vescovi and Jaci VanHeest

The Science of Judo
Edited by Mike Callan

Modelling and Simulation in Sport and Exercise
Edited by Arnold Baca and Jürgen Perl

Modelling and Simulation in Sport and Exercise

Edited by Arnold Baca and Jürgen Perl

Routledge
Taylor & Francis Group
LONDON AND NEW YORK

First published 2019
by Routledge
2 Park Square, Milton Park, Abingdon, Oxon OX14 4RN

and by Routledge
711 Third Avenue, New York, NY 10017

Routledge is an imprint of the Taylor & Francis Group, an informa business

© 2019 selection and editorial matter, Arnold Baca and Jürgen Perl; individual chapters, the contributors

The right of Arnold Baca and Jürgen Perl to be identified as the authors of the editorial material, and of the authors for their individual chapters, has been asserted in accordance with sections 77 and 78 of the Copyright, Designs and Patents Act 1988.

All rights reserved. No part of this book may be reprinted or reproduced or utilised in any form or by any electronic, mechanical, or other means, now known or hereafter invented, including photocopying and recording, or in any information storage or retrieval system, without permission in writing from the publishers.

Trademark notice: Product or corporate names may be trademarks or registered trademarks, and are used only for identification and explanation without intent to infringe.

British Library Cataloguing-in-Publication Data
A catalogue record for this book is available from the British Library

Library of Congress Cataloging-in-Publication Data
A catalog record has been requested for this book

ISBN: 978-1-138-05993-1 (hbk)
ISBN: 978-1-315-16329-1 (ebk)

Typeset in Galliard
by Out of House Publishing

Contents

List of figures vii
List of tables xi
About the contributors xii
Preface xv

PART I
Aspects of human and sports motion 1

1 Motor activity: motor control and learning 3
  JOSEF WIEMEYER

2 Muscle mechanics 22
  SIGRID THALLER AND HARALD PENASSO

3 Rowing 50
  MICHAŁ WYCHOWAŃSKI AND ARNOLD BACA

PART II
Team sports and the modelling of playing processes and tactical behavior 71

4 Soccer: process and interaction 73
  JÜRGEN PERL AND DANIEL MEMMERT

5 Modelling in the analysis of tactical behavior in team handball 95
  MARKUS TILP

6 Basketball 108
  JAIME SAMPAIO, BRUNO GONÇALVES, NUNO MATEUS,
  ZHANG SHAOLIANG AND NUNO LEITE

PART III
Evaluation concepts and techniques 127

7 Tournaments 129
   PETER O'DONOGHUE

8 Key performance indicators 146
   JÜRGEN PERL AND DANIEL MEMMERT

PART IV
Physiological conditions of being successful 167

9 Marathon 169
   STEFAN ENDLER

10 Training 187
   CHRISTIAN RASCHE AND MARK PFEIFFER

PART V
Tools and equipment used in sport 209

11 Modelling and simulation to prevent overloads in
   snowboarding 211
   VEIT SENNER, STEFAN LEHNER, FRANK I. MICHEL
   AND OTHMAR BRÜGGER

12 Methods to gather key performance indicators for
   prosthetic feet 237
   FELIX STARKER, ERIC NICKEL AND ANDREW HANSEN

   Index 267

Figures

1.1	Classification scheme for sensori-motor activities	4
1.2	Illustration of the javelin throw	6
1.3	Overview and examples of qualitative and quantitative models of motor activities	8
1.4	Examples of two Fuzzy sets for rock and roll	9
1.5	Illustration of a two-dimensional 4 x 4 KFM	11
1.6	Arbitrary task space for an aimed throwing movement	13
1.7	Internal models	13
1.8	Illustration of a countermovement jump (CMJ)	17
1.9	Vertical forces of a CMJ	18
2.1	Schematic representation of a sarcomere	26
2.2	Sliding of the actin and myosin filaments against each other leads to a different overlap at lengthening, at optimal length and at shortening	26
2.3	The three different models of the force-length relation in the muscle show the isometric muscle force normalized to the isometric force at optimal length	28
2.4	Hyperbolic shape of the force-velocity relation with asymptotes	32
2.5	Eccentric and concentric force-velocity relations	34
2.6	Two commonly used combinations of a contractile element together with parallel and serial elastic elements	37
2.7	Knee extension movement on a leg press	38
2.8	Individual force-velocity relations using identified parameter values in Hill's equation	40
2.9	Sketch of the situation in example 2	44
3.1	Main training factors and environmental conditions affecting the result in rowing	52
3.2	Stages of studying the dynamics of the system	58
3.3	Statistical classification from sensor data	59
3.4	Physical model of the athlete-boat-oars system	61
3.5	An example of single oar force and total driving force in the first 10 s of a regatta	64

viii *Figures*

3.6	An example of boat acceleration and boat velocity in the first 10 s of a regatta	64
4.1	Position, formation and process on a coach's black board	79
4.2	The geometric relations between player positions build a formation of the players, the similarity of which can be recognized by means of a respectively trained network	80
4.3	Examples of defense and offense formation types together in the order of their frequencies	81
4.4	Coincidence matrix of defense formation types against offense formation types	81
4.5	Success matrix of the coincidence matrix from Figure 4.4	82
4.6	Reduced success matrix containing only successful interactions of team B	82
4.7	Voronoi cells of the players of team A and team B	84
4.8	Space-control profiles of team A and team B	85
4.9	A pass from player A1 to player A2, passing three opposition players	85
4.10	Pass efficiency depending on the number of outplayed players and the distances to closest opposing players at the moments of passing and receiving	86
4.11	Three levels of modelling	91
5.1	Schematic display of ball path in the last five passes prior to a shot at goal	97
5.2	Data coordinates of player positions as input for an ANN	98
5.3	Computer positions, camera positions and camera perspectives during the video recording of handball games	99
5.4	MASA user interface	101
5.5	Illustration of the eight most common offensive patterns	102
5.6	Combination of one specific offensive pattern against different defensive patterns	103
5.7	Mean areas around actual shot position based on mean distances between predicted and actual shot position of the two approaches, respectively	105
6.1	Depiction of the three steps of a possible approach	112
6.2	Histograms of teams' constitution profile according to high- and low-scoring players	117
6.3	Teams' constitution profile according to high- and low-scoring players for each playing position	118
6.4	Lower-dimensional projection of the predictor space containing six predictors	119
6.5	Target values by predictors for initial focal cases and nearest neighbors	120
6.6	Lower-dimensional projection of the predictor space containing nine predictors	121
6.7	Target values by predictors for initial focal cases and nearest neighbors	122

6.8	Lower-dimensional projection of the predictor space containing seven predictors	123
6.9	Target values by predictors for initial focal cases and nearest neighbors	124
7.1	Model using regression analysis	133
7.2	An example of heteroscedasticity within data	134
7.3	Structure of the UEFA Nations League	139
7.4	Simulator design	140
8.1	Fuzzy quality classes regarding the number of passes	151
8.2	Trained neural network with numbered squares representing neurons	153
8.3	Attribute table comparing the attributes of a selected discipline and the most similar prototype together with the respective differences	154
8.4	Voronoi cells of attacking and defending players with the numbers of the respective players	157
8.5	Time-diagram of a play showing profiles of space-control rates in the opponent's areas and ball-control events	157
8.6	Time-diagram showing the success values of team A and team B calculated for the interval of the respective last 300 seconds	158
8.7	Interactive coupling of the time-diagram and animation or video components	161
8.8	Examples of success values of team A and team B and the advantages of team A depending on respective exploiting rates	163
8.9	Diagram showing the values of AdvA depending on ERA and ERB	164
8.10	Excel diagram showing SVA, SVB and the advantage for A	165
9.1	Keul's (1979) model	175
9.2	Schematic diagram of speed determination using GPS	177
9.3	Structure of the performance potential (PerPot) model	178
9.4	HR modelling with different smoothing factors	182
9.5	HR modelling using smoothing factors out of the range	183
10.1	Overview of existing performance models	189
10.2	Type B performance model	195
10.3	15 weeks of training and performance measurements for an elite rower	202
10.4	Performance simulation with the IRmod Model	203
11.1	Backward fall starting condition: joint angles and joint-center location	213
11.2	Overview of the chosen modelling concept	215
11.3	Backward falls simulated with MBS models	216
11.4	Detailed MBS model of the right upper extremity	217
11.5a	Wrist splint models	219
11.5b	Fixation of dorsal splint to the forearm	219
11.6	Zones and bone tissues distinguished in our FEA model of the radius	222

x Figures

11.7a	Relative reductions in wrist extension when different protector designs are compared to the reference model without a protector	226
11.7b	Relative reductions in axial wrist joint force for different protector designs compared to the reference model without a protector	226
11.8	The adult group's FEAs of stress distribution in the radius bone	228
11.9	FEAs of the stress distribution at the radius in the adolescent model when wearing the long dorsal protector	229
11.10	FEAs of the stress distribution at the radius in the child model when wearing the long dorsal protector	230
11.11	Displacement of the dorsal protector at impact	232
12.1	Input and output parameters describing a prosthetic foot	239
12.2	Different coordinate systems based on the location to measure data	241
12.3	Translating the human walking motion in a simplified machine setup where the ground articulates to the foot	242
12.4	Setup of the prosthetic foot in a machine according to DIN (2016)	243
12.5	Displacement and rotation occurring at the prosthetic foot	244
12.6	Force curves of two test samples	246
12.7	The highest moment of the prosthetic foot occurs in the sagittal plane	247
12.8	The effective lever arm not only differs in maximum values between different feet but also describes the time and duration of transition from heel to midfoot, full ground contact and, finally, heel-off	248
12.9	The shape of the effective rocker can be simplified as a circle radius during the support stance phase	250
12.10	Measuring quasi-stiffness curves from single loading tests reveals the energy storage and return without clarifying whether the motion occurred in a deformation or a rotation	251
12.11	Based on each prosthetic foot's design, the ankle angle vs. ankle moment not only represents the range of motion it undergoes during a full roll-over under a standardized input angle slope, but also how much energy during this rotation is dissipated	253
12.12	Dimensions of a prosthetic foot used for the example	254
12.13	Schematic of a drop-test setup	256
12.14	Linear spring-mass-damper diagram	257
12.15	Schematic of an example for downward jumping	257
12.16	Impact displacement and vertical ground reaction force for the prosthetic foot model	260
12.17	Impact forces vs. time of a drop test of a prosthetic foot	261
12.18	First impact force vs. time of a prosthetic foot	262
12.19	Spring mass oscillations (later peaks of impact data)	262
12.20	Plot of oscillation peaks with exponential fit curve	263

Tables

6.1	Linear classification function coefficients for higher- and lower-scoring players and structure coefficients for each position	115
7.1	Standard deviation of absolute residual values	137
7.2	Chances of team success in the UEFA Nations League 2018–2019	142
7.3	Progression statistics produced by the Euro 2020 simulator	143
8.1	Some examples of KPIs on increasing levels of complexity	148
8.2	Some examples of player- and team-oriented KPIs	155
9.1	Prediction of marathon goal times using Riegel's formula	172
9.2	Actual times for three athletes at three competitions	172
10.1	Exemplary delay values, decay/delay factors and decay vs. delay factor % deviations	198
10.2	Included parameters of selected performance models in the optimization process	199
10.3	Exemplary data and parameter set to use a performance model by hand	204
11.1	Anthropometric data for modeled age groups from the Size Germany database	215
11.2	Overview of modeled wristguard concepts and versions	219
11.3	Young's modulus for the different materials	221
11.4	Maximum load and extension of the wrist joint during impact	225
11.5	Maximum shear stresses in the adult model in the reference condition and when wearing each of the three versions of the dorsal and palmar protectors	227
11.6	Stress reduction relative to the reference model for the adult, adolescent and boy wearing the long versions of the dorsal and palmar protectors	229
12.1	Specific parameters extracted from test data presented earlier	254

About the contributors

Arnold Baca
Department of Biomechanics, Kinesiology and Applied Computer Science
University of Vienna
1150 Vienna, Austria

Othmar Brügger
Bfu – Swiss Council for Accident Prevention
3011 Berne, Switzerland

Stefan Endler
Computer Science in Sport
University of Mainz
55122 Mainz, Germany

Bruno Gonçalves
Research Center in Sports Sciences, Health Sciences and Human Development (CIDESD) CreativeLab Research Community
University of Trás-os-Montes e Alto Douro
Vila Real, Portugal

Andrew Hansen
Minneapolis VA Health Care System, Minneapolis Adaptive Design & Engineering (MADE) Program
University of Minnesota, Department of Rehabilitation Medicine
One Veterans Drive (Research – 151)
Minneapolis, MN 55417, USA

Stefan Lehner
Department Sport Equipment and Material, Institute of Ergonomics
Technical University of Munich
85747 Garching, Germany

Nuno Leite
Research Center in Sports Sciences, Health Sciences and Human Development (CIDESD) CreativeLab Research Community
University of Trás-os-Montes e Alto Douro
Vila Real, Portugal

Nuno Mateus
Research Center in Sports Sciences, Health Sciences and Human Development (CIDESD) CreativeLab Research Community
University of Trás-os-Montes e Alto Douro
Vila Real, Portugal

Daniel Memmert
Institute of Training Sciences and Sport Informatics
German Sport University Cologne
50933 Cologne, Germany

Frank I. Michel
SCM – Sports Consulting Michel
88085 Langenargen (Bodensee), Germany

Eric Nickel
Minneapolis VA Health Care System, Minneapolis Adaptive Design & Engineering (MADE) Program
One Veterans Drive (Research – 151)
Minneapolis, MN 55417, USA

Peter O'Donoghue
Cardiff School of Sport
Cardiff Metropolitan University, Cyncoed Campus
Cardiff, CF23 6XD, UK

Harald Penasso
Institute of Sports Science
University of Graz
8010 Graz, Austria

Jürgen Perl
Computer Science in Sport
University of Mainz
55122 Mainz, Germany

Mark Pfeiffer
Institute of Sport Science
University of Mainz
55122 Mainz, Germany

Christian Rasche
Institute of Sport Science
University of Mainz
55122 Mainz, Germany

xiv *About the contributors*

Jaime Sampaio
Research Center in Sports Sciences, Health Sciences and Human Development (CIDESD)
CreativeLab Research Community
University of Trás-os-Montes e Alto Douro
Vila Real, Portugal

Veit Senner
Sport Equipment and Materials
Technical University of Munich
85747 Garching, Germany

Zhang Shaoliang
Faculty of Physical Activity and Sports Sciences
Polytechnic University of Madrid
Madrid, Spain

Felix Starker
Össur hf., Grjóthalsi 5
110 Reykjavík, Iceland

Sigrid Thaller
Institute for Sport Science
University of Graz
8010 Graz, Austria

Markus Tilp
Institute for Sport Science
University of Graz
8010 Graz, Austria

Josef Wiemeyer
Institute of Sport Science
Technische Universität Darmstadt
64289 Darmstadt, Germany

Michał Wychowański
Department of Rehabilitation
Józef Piłsudski University of Physical Education in Warsaw
00–968 Warsaw, Poland

Preface

Roughly speaking, modelling means a mapping from one system, its components and dynamics to another system, its components and dynamics.

As a basic example, one can take the human's perception of the surrounding world:

The sensory organs acquire data from the surrounding environment and transfer it to the neurons of the brain. First, a mapping picture is generated by encoding the incoming signals into neural activity, enabling to understand what is going on. Next, based on accumulated experience, these pictures are analyzed and interpreted, generating a view on dynamic processes and the reasons for them, enabling the brain to understand why things are happening.

This very simple example clarifies and separates the two most important and quite different aspects of modelling:

One aspect is the description of the system's behavior, which can be found predominantly in statistical descriptions of distributions or in mathematical interpolations of discrete values. In what follows, such models are called "descriptive."

The other aspect is the causality analysis, where the dynamics of components' interactions are described—for instance, mapped to equations—which can be found in system dynamics approaches or in differential equations analyses. In what follows, such models are called "analytic."

In both cases, one main idea is not only to understand a system's state or behavior but also to calculate its changing states or developing behavior, i.e., to simulate the system by means of the model.

One quite simple but instructive example for distinguishing between these two aspects is that of the body mass index (BMI).

In the 1970s, the maximum life expectancy was correlated to a BMI of 23.7. This led the World Health Organization to strongly recommend a maximum value of 25.0 for a healthy life, whereas a BMI > 25 indicated corpulence and was said to reduce life expectancy. Therefore, the BMI value was intended to give human beings an orientation for optimizing their nutrition. In terms of modelling, this means that the BMI approach simulates the relationship between body weight and life quality and expectancy.

Meanwhile, analyses show that the optimal BMI value has changed—24.6 in the 1990s to 27.0 nowadays—but no explanation for this change can be found.

Obviously, the problem of the BMI approach is that the model is descriptive and not based on causality analysis. Therefore, it does not allow for any simulative prediction of life expectancy of individuals.

In other words, the crucial fact is that the BMI model only meets one of the two main tasks of a model, namely the diagnostics of the system's state, while the other main task, the prognosis of the system's future behavior, is missing. As often happens in similar situations, the consequence is that the model as a whole is not suitable because it was originally seen as an orientation for weight-watching over a long timescale.

In practice, the choice between either descriptive or analytic modelling depends on the complexity of the system to be modeled, on any claims of model precision and on whether the effort is justifiable and/or sufficient.

While the BMI example deals with humans' internal dynamics, the following example focuses on the relation between humans and the universe: If two lovers are sitting on a garden bench on a mild summer's night, the stars in the sky are simply wonderful. No precision of description is necessary to feel the greatness of the universe and of the moment; scientific descriptions would more likely destroy the romantic atmosphere. If, however, we are to interpret stars and stellar constellations as mighty beings, as our ancestors did, then it is necessary to describe and predict their behavior. Enormous effort seems reasonable in attempting to model such a system of influences precisely, as is apparent from all the activity at Stonehenge, the famous monument in England. Nowadays, the focus is on the astrophysical claim for understanding the complex dynamics of the universe and for predicting its future behavior as precisely as possible. To this end, almost unlimited effort seems to be justifiable, like large radio telescope fields or satellite-based telescopes.

In sports, things are a bit different, but comparably complex and difficult. The reasons for this complexity are the largely differing types of interaction for one athlete: components within the athlete's own body, with other athletes and with sports equipment and technical tools. In contrast to modelling technical systems, behavioral modelling plays an important role when modelling sports. The following three examples try to point out the challenges of behavioral modelling in sports:

In the case of the shot put, the system to be modeled consists of two components with their specific dynamics and the interaction or transfer between them: The athlete and the shot. The dynamics of the shot can easily be described by physical models. More difficult is the interaction, i.e., the power transfer from the athlete to the shot. Although the optimal angles of the arm and the hand can be calculated by physical laws as well, the problem is that the athlete has to follow the laws as precisely as possible. Therefore, modelling the athlete's behavior is the biggest challenge: Extremely complex biomechanical dynamics prevent a direct transfer from technical data to the athlete's motion. Instead, long-term training is necessary to develop motion patterns that help the athlete to feel and control the correctness of their shot put technique.

Quite differently, a marathon runner fights against him- or herself for more than 2 hours. It is not the technical equipment that has to be modeled—disregarding non-optimal shoes or clothing—but rather the effect of physical load on the changing physiological state of the runner. Nevertheless, high precision of modelling is necessary for optimizing the physical load, as even minimally exceeding the optimal speed at any point of the run can cause a sudden breakdown later on. Therefore, a predictive dynamical systems model is required to find the optimal speed control for the runner.

Finally, in a game like soccer—although physiological states of the players are significant—the focus is on the interaction and processes between players. Before being able to start data-based modelling, simply recording data is extremely difficult. On the one hand, recording technical data like players and the ball's positions is a challenge. Only in about the last 5 years has the situation improved significantly by automatic recording of players' positions. On the other hand, there is no way to distinguish between intention and behavior. If a pass from player A to player B fails, there are several possible reasons like "A passed in the wrong direction" or "B moved to the wrong position." In order to optimize passing by modelling, analyses and training, intention and behavior have to be known and to be matched by means of technical and tactical exercise. Obviously, the situation is even more difficult if not just one pass but processes are of interest, such as sequences of passes in the context of tactical situations and in interaction with the opposition's activities.

So far, only the pair "descriptive vs. analytic" has been used to classify the abilities of the model: Descriptive modelling uses observable items for a description of present behavior, while analytic modelling uses internal dynamics for a prediction of future behavior. Originally, under the aspects of diagnosis and prognosis, a model should meet both tasks. However, as the examples demonstrate, this is not always that easy in practice, hence many models are reduced to the descriptive part, particularly in sports.

In addition to the pair "descriptive vs. analytic," there are several pairs of classifying properties that span a diversity of model types when combined differently. The following table presents a brief overview of model types with some examples from sports.

Type of	*Focus*	*Examples*
understanding		
white box	internal dynamics	basic physiological processes
black box	external behavior	immediate response to load
observable items		
quantitative	technical data	number of passed players
qualitative	behavioral information	choice of directed player
process resolution		
discrete	time step oriented	key steps of a player's attack
continuous	time flow oriented	trajectory of a player
process description		
state-event equations	context dependent, following local rules under changing conditions	motion with compensation effects
differential equations	context independent, following global rules under constant conditions	idealized motion

It would be an instructive exercise to type the models presented in the following articles under the classifying items from above which, incidentally, is not at all unambiguous and depends strongly on one's point of view and the focus of modelling.

From a bird's-eye view, the peloton in a race like the Tour de France looks like a continuous stream, the movement of which is determined only by course data:

- Only external behavior can be observed (black box).
- Technical data like speed and acceleration can be measured (quantitative).
- The process resolution is time flow oriented (continuous).
- Based on the course data, contexts are not important over a short distance and conditions are constant (differential equation).

From an individual rider's point of view, the model may look quite different:

- Data about speed, power and heart rate give the rider information about their internal dynamics (white box).
- Based on experience, the rider knows when they're ready to launch a surprise attack (qualitative).
- At the right moment, as soon as a gap appears, the rider will speed up (discrete).
- The rider switches behavior from one moment to the next, following changing contexts and conditions (state-event).

Of course, the individual chapters in this book cannot give a complete list of modelling and simulation or types of models in sports. Instead, the articles and presented examples are oriented in present activities in sports. They reflect the different approaches with their different concepts and methods and are grouped into five main topics:

Part I (chapters 1–3) deals with aspects of human and sports motions. It covers various issues from biomechanics, which can be said to be the oldest and most developed discipline utilizing modelling in sport, as physics and sport have been developing a fruitful partnership since the 1950s.

In the first chapter, addressing motor activity, Wiemeyer gives an introduction to the concepts and techniques of motor control and motor learning, where "motor control denotes the capability to successfully perform goal-directed movements, whereas motor learning is defined as an experience-based process leading to a more or less permanent change of perception and action" (p. 5). The perspectives of modelling stretch from biomechanical and neurophysiologic to psychological approaches, generating a great diversity of model types.

In particular, Wiemeyer discusses different types of motor activity modelling, thereby giving an informative introduction to the practical meaning of the underlying modelling paradigms.

Aspects of biomechanical modelling in general and modelling muscle mechanics in particular are discussed in the chapter by Thaller and Penasso. Three goals are highlighted. First, models relying on movement-independent parameters are derived. Then, methods for individual model parameter estimation are presented. Finally, the question of sensitivity of models to their parameters is assessed. Altogether, valuable insights in the process of setting up appropriate individual models and the problems involved are given.

Completely different modelling approaches for answering practical issues in the sport of rowing are presented in the third chapter by Wychowański and Baca. On the one hand, methods from classical Newtonian mechanics are applied; on the other, the potential of data-driven models is assessed. Moreover, it is demonstrated that different factors have to be considered when assessing performance in a particular sport. In the case of rowing, the outcome of a competition is not only affected by the motion technique, which is the main focus of the chapter, but also by the athletes' physical features and the tactical behavior.

Part II (chapters 4–6) considers team sports and the Modelling of playing processes and tactical behavior. modelling team sports is a very difficult and complex task. Although the first attempts were made in the 1950s, those early approaches were rather theoretic because of a lack of data-recording techniques and, therefore, useful data. Only since about 2010, data from team sports—such as position of players and the ball—have become available and have helped to make team game modelling useful.

In Chapter 4, Perl and Memmert discuss the problem of modelling complex behavior and its evaluation in the interaction of teams of players. Together with the need to reduce reality—caused by a lack of usable data as well as by the restrictions of simplifying models—this leads to two central questions:

> - How precisely can an intentionally characterized dynamic system like a play be modeled in the first place?
> - What are the consequences of modelling and analysis errors for results and evaluations?
>
> (p. 74)

Some approaches dealing with position data of the players and the ball are introduced, and, among other things, it is shown how complex team interaction can be reduced to the interaction of simple patterns by means of artificial neural networks.

This approach of net-based pattern analysis is dealt with in more detail in the subsequent chapter, where Tilp models and analyzes sequences of actions in handball. After introducing a number of up-to-date approaches, he describes an attempt of recognizing and characterizing offensive playing processes based on self-organizing neural networks. One promising finding of his analyses is that:

> In this approach, the deviation of only 1.2 m demonstrated a promising accordance of single action sequences with the related patterns. One

important advantage of this approach compared to classical approaches is that context information is taken into account. Even when the final shot position of two action sequences is the same, the preceding actions might have been significantly different and lead to different defensive countermeasures.

(p. 101)

Different and more difficult is the situation in basketball, as Sampaio, Gonçalves, Mateus, Shaoliang and Leite explain in Chapter 6: "The studies focused on positional-derived variables in basketball are still very limited to small samples ..." (p. 110).

Usually, box-score statistics are used for explaining the game dynamics and outcome in basketball and therefore lead to the problems that are discussed in Part III. As in soccer and handball, however, position data of the players and the ball become increasingly important in basketball:

Player-tracking technology is one of the most recent technological advances in basketball. Powerful computer vision systems were designed with fine-tuned algorithms capable of tracking with relatively high accuracy the players' positioning and, subsequently, all derived variables such as distance covered and speed.

(p. 110)

Evaluation concepts and techniques, in particular in the context of sports, are the focus of Part III (chapters 7–8). Whether an athlete, a player or a team in a competition is as successful as expected determines many aspects: grouping for tournaments, ranking for qualifying or just money for positioning. Measuring or evaluating success, however, is a difficult task if the aim is to reduce complex activity to just one ranking number. Therefore, a lot of quite different methods and approaches have been developed.

In Chapter 7, O'Donoghue introduces the problems of designing tournaments, which not only depends on the performance and qualification of the involved athletes or teams but also:

Sports tournaments can be organized as round-robin leagues, knockout competitions or a combination of both. There may be qualifying stages required due to restrictions on the number of participating teams or players. Tournament structures have changed to improve participation, revenues, entertainment value and safety.

(p. 129)

O'Donoghue discusses types of performance indicators, their meaning for tournament design and how modelling and simulation of tournament dynamics can help to understand a tournament's dynamic behavior and improve the adequacy of its design.

He presents the main algorithms and data structures that are used to produce useful output information that is ultimately required by decision makers.

Methods and indicators as dealt with in O'Donoghue's chapter are also necessary for evaluating individual performance and success of players and teams. In Chapter 8, Perl and Memmert start by discussing the meaning and usefulness of simple indicators like the number of passes or percentage of ball possession. Such indicators can be understood as a radical model-oriented reduction—often, incidentally, reducing or even erasing useful information in the process. From a modelling viewpoint, approaches are introduced that try to combine necessary reduction with protection of key information in order to indicate useful performance aspects. However, the challenge remains in finding the right balance between the precision of analysis results and identifying the best way to transfer interpretation of those results to the practice of playing in order to improve performance.

The physiologic conditions of being successful, i.e., strategies of training and competition, are assessed in Part IV (chapters 9–10). The physiologic resources of an athlete are limited; they can be improved by intelligent and adequate training or diminished by bad training strategies or overtraining. In competitions—in particular in endurance sports, but also in longer-lasting games like soccer or tennis—an optimal progress of physical exertion is of central importance.

If, for example, a marathon runner reaches the finish comparatively fresh and relaxed but seconds over their personal best time, they will not really be satisfied. This is one problem Endler addresses in his chapter on marathon, i.e., how to achieve best performance without exceeding physiologic limit. This means that:

> For an athlete, it is helpful to know an approximated finish time to avoid exhausting their energy reserves and face an early breakdown. Knowing an optimal marathon finish time can be used to prescribe an individual pacing strategy. Several different techniques can be used for that approximation. [...] Other techniques use load parameters, e.g., running speed in km/h or min/km and parameters measuring reaction of the body due to load, e.g., heart rate (HR), lactate or oxygen uptake.
>
> (p. 169)

In particular, Endler presents one model that describes the dynamics of performance and fatigue and demonstrates how it can be used to control running speed, depending on continuously measured heart rate.

Even more than in a competition like marathon, an overload in one phase of the training can cause a delayed breakdown days or even weeks later. Consequently:

> [...] the effect of training on performance is of vital interest to coaches organizing and scheduling their athletes' training programs. Therefore, it is important to learn the mechanics and basics of how training load transfers toward adaptation processes and ultimately affects performance.
>
> (Rasche & Pfeiffer, p. 187 this book)

xxii *Preface*

In Chapter 10, Rasche and Pfeiffer introduce a number of appropriate models and discuss their advantages and disadvantages, depending on the field of application.

Finally, Part V (chapters 11–12) deals with the tools and equipment used in sport and that are therefore subject to design and optimization. There is a broad spectrum of such tools, ranging from bobsleighs more expensive than luxury cars to shoes tested in special laboratories for optimal adaption to special moving conditions. Moreover, particular tools like prostheses can help not only to improve movement but also to enable movement at all.

Different types of models are combined by Senner, Lehner, Michel and Brügger in order to answer a specific research question in the area of prevention of overloads in snow sports: "Which type of design or which functional elements of snowboard wrist protectors can reduce stress in the wrist and forearm bones in a typical backward fall situation, in particular for the target group children and adolescents?" (p. 214).

Again, biomechanical models are the focus of this chapter. In particular, various computer-aided engineering tools are introduced and discussed.

A fascinating final chapter on the design of protheses is presented by Starker, Nickel and Hansen. It has always been a big challenge to replace limbs with mechanical tools. Contemporary prostheses show an amazing combination of different technology from sport and healthcare as well as from mechanical and electronic engineering. The amount of research activity in this field is a striking indicator of the increasing significance of prostheses. Not least, and this may close the circle to the field of motor activity from the first chapter, prostheses have helped to develop a quite new field of sports for handicapped people, significant of course for individuals in particular and for the Paralympic Games in general.

We would like to acknowledge all the authors for their interesting contributions which together provide a fascinating insight to the world of modelling and simulation in sport and exercise.

Last but not least, our warmest thanks to Martin Gröber for assisting us in reaching this difficult and challenging goal.

Part I
Aspects of human and sports motion

1 Motor activity
Motor control and learning

Josef Wiemeyer

Introduction

Humans perform movements or sensori-motor activities to interact with their environment in order to reach certain goals. Among others, this can be to arrive at a location (e.g., restaurant), to communicate (by speech, gestures and mimics) or to reach for and catch an object (e.g., a basketball). Every goal-directed movement depends on a systematic coupling of perception and action. Perception provides information and feedback delivered by the senses for the adequate control of action. However, there is no generally agreed classification or taxonomy of human sensori-motor activity (HMA). In sport, sensori-motor activities are classified according to a qualitative model into generic activities, termed *motor abilities* and specific activities comprising characteristic features, called *motor skills* (see Figure 1.1). Motor abilities can be further subdivided into *condition(ing)*, i.e., energetic factors, and *coordination*, i.e., informational factors. Motor skills can be differentiated depending on the particular context, e.g., vocation, leisure or sports.

As stated in the introduction to this book, HMA are complex phenomena. Modelling approaches cover a wide variety of types, including qualitative and quantitative models as well as white box and black box or discrete and continuous models. One approach claims that HMA are constrained by numerous factors, including the individuum (e.g., skill and conditioning level), the environment (e.g., slope or weight and shape of sports equipment) and the task (e.g., moving under time or precision pressure). Concerning the individuum, two important biomechanical structures constrain movement: joints and muscles. On the one hand, joints are characterized by a particular structure of bones, cartilage, capsule and ligaments constituting the joint workspace; on the other hand, muscles act on the joints (under the control of the central nervous system) to exploit the joint workspace. For example, the shoulder joint is a connection of the scapula and the humerus; it has a loose capsule and only weak ligaments forming a ball joint and allowing for 3 degrees of freedom (DOF): abduction–adduction; anteversion–retroversion; supination–pronation (internal or external rotation). Regarding the sum of DOF in the human movement system, 147 joints offer about 238 DOF (Saziorski, Aruin & Selujanow, 1984) and the number of muscles exceeds 500.

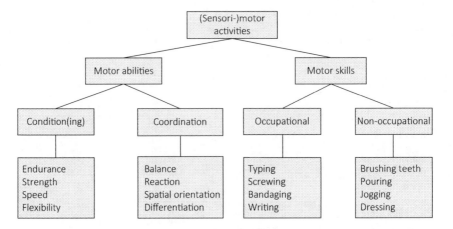

Figure 1.1 Classification scheme for sensori-motor activities.

Therefore, human joints and muscles offer much more DOF than are needed for a unique solution of a motor problem, e.g., reaching or throwing. This issue is termed the "DOF" or "Bernstein's Problem" (e.g., Bernstein, 1967; Whiting, 1984; Newell & Vaillancourt, 2001) or solution manifold (e.g., Müller & Sternad, 2009). On the one hand, Bernstein's Problem implies numerous options for flexible control of acquired motor skills (solution manifold or solution space); on the other hand, it illustrates the difficulties of motor learning. Particularly in early stages of the acquisition process, learners have to find an appropriate and stable solution within the solution space. The DOF or Bernstein's Problem is one of the fundamental issues of motor control and motor learning. There are different approaches to this problem. For example, motor approaches claim that internal representations like motor programs, schemata (e.g., Schmidt, 1980, 2003; Summers & Anson, 2009) or internal models comprising forward and inverse models (e.g., Wolpert, Miall & Kawato, 1998; Kawato, 1999; Maurer, Maurer & Müller, 2015; Argyropoulos, 2016) play an important role in motor control and learning. In contrast, dynamical systems approaches claim that order and disorder of human movements emerge from "self-organizing" dynamic interactions of the human system and the environment (e.g., Beek, Peper & Stegeman, 1995; Kelso, 1997). As a consequence, quite different types of modelling are used by the two approaches: While motor approaches use mainly computational approaches of internal representations, dynamical systems approaches prefer computational approaches of system dynamics, i.e., the interaction of internal and external constraints influencing the stability and variability of movements.

Regarding motor learning, the "Bernstein hypothesis" (Bernstein, 1967; Schneider, Zernicke, Schmidt & Hart, 1989; Vereijken, Emmerik, Whiting & Newell, 1992; critical discussion: Newell & Vaillancourt, 2001; Latash, 2010) claims that the motor system passes through specific stages in the process of acquiring new movements: First, DOF are actively blocked ("freezing") to simplify the control issue. Subsequently, the DOF are gradually released ("freeing") until, finally, the DOF and their dynamic interplay are systematically exploited ("exploitation").

In this chapter, *motor control* denotes the capability to successfully perform goal-directed movements, whereas *motor learning* is defined as an experience-based process leading to a more or less permanent change of perception and action.

Motor learning and motor control can be analyzed and modeled from different perspectives. Among the most prominent perspectives are the biomechanical, the neurophysiological and the psychological perspectives. As a consequence, (computational) models of the process and results of motor control and motor learning also show a great diversity, ranging from explicit or white-box models like general motor programs (e.g., Schmidt, 1980) to implicit or black-box approaches like artificial neural networks (ANNs), genetic algorithms and deep-learning approaches.

In 1982, Marr introduced an important distinction between three levels of understanding information processing systems: a computational, an algorithmic or representational and an implementational level. At the *computational level*, questions concerning the goal and appropriateness of the computation as well as the logic of the implementation strategy have to be answered. At the *algorithmic or representational level*, the appropriate representation of input and output as well as the input-output transformation have to be specified. At the *implementational level*, the question of physical (or biological) implementation has to be solved. Of course, it is debatable whether a bottom-up approach, i.e., starting at the implementational level, or a top-down approach, i.e., starting at the computational level, is more adequate for modelling and simulation. Regardless of starting point, a valid modelling and simulation of human motor activities has to consider these three levels; for example, it does not make sense to think about the computational level of motor control without considering the implementational level including certain brain structures as well as peripheral neurons, sensors, muscles and joints. On the other hand, transferring principles of human motor learning to robot learning has to consider the implementational differences (actuators, sensors and control loops) between humans and robots.

This chapter is structured as follows: First, the concept of HMA is further analyzed in three aspects (structure, control and learning). Second, selected concepts and methods of modelling are introduced and described. Third, the advantages and disadvantages as well as the current state of probation, confirmation and acceptance are analyzed. Fourth, the results of modelling HMA by applying different models are summarized. Finally, examples and exercises are presented to actively work on modelling HMA.

Project/problem

Modelling and simulating HMA, i.e., motor control or motor learning, as well as the structure of motor skills, is an interdisciplinary enterprise that has to consider the different perspectives mentioned in the introduction. First, motor activities result in movements that can be observed from the outside perspective. Therefore, qualitative and quantitative data about movement outcomes can be used to describe and explain motor activities. Second, the process of motor control and motor learning can be modeled using computational models.

In this section, the modelling of movements will be addressed first, followed by motor control and motor learning. Three questions will be discussed:

(1) How can HMA be structured?
(2) "Bernstein's Problem": How can the complexity of the system (joints, muscles, motor units) be reduced to a controllable level?
(3) How does the sensori-motor system acquire control (learning)?

Structure of human sensori-motor activities

In principle, HMA are a more or less complex combination of constitutive parts. Taking the example of a throwing movement, specific movements of the arms, the trunk and the legs contribute to the solution (see Figure 1.2). The throwing arm performs an initial backward movement, followed by a forward movement. The legs perform the initial run-up, followed by a transition to the throwing stance. The trunk, and particularly the shoulder and the pelvis, initially rotates backward, followed by a forward movement. A successful throw shows a characteristic structure of movements being performed either simultaneously or successively. In sport, throwing skills are often required. For example, in track and field athletics, various disciplines include the task of throwing a device as far as possible, e.g., a javelin, a discus or a hammer. Here, height, angle and velocity of release as well as aerodynamics determine how far the device will travel. Furthermore, in many sports, the goal is to hit or approach a target or a target zone, e.g., a basket, a dartboard, a teammate or a goal. Therefore, precision (and often speed) of the movement is important. As will be shown in the following section, aimed throwing movement can be comprehensively modeled using biomechanical approaches.

Structuring (complex) motor activities poses a lot of problems:

- What is the criteria for identifying parts of the movement?
- Which movements are important and which movements are less important for achieving the desired goal?

Figure 1.2 Illustration of the javelin throw.

- How is the outcome affected by the variation of movements?
- Is there an ideal movement structure generally valid for all persons?

There are numerous ways to structure movement (overview: Wiemeyer, 2003). These models will be addressed in the next section.

Motor control: "Bernstein's Problem"

As has been pointed out in the introduction, the core problem of motor control is to turn the various and redundant DOF into a controllable system. Several systems are involved in the control of motor activities:

- The biomechanical system, comprising joints, muscles and tendons;
- The neurophysiological system, involving brain, spinal cord, sensors and neuromuscular links;
- The psychological system, comprising perception, action, decision-making, cognition, emotion, volition and motivation;
- The metabolic and cardiorespiratory systems providing energy for the respective motor activities.

As will be shown in the next section, there are different strategies to control movement, for example coupling DOF or exploiting flexible solutions. It does not make sense for the control center to try to control each movement feature, since this would raise many control issues, e.g., the non-determinant relation of central impulse (e.g., motor commands) and peripheral response (e.g., forces and torques).

Motor learning: acquiring control over movement

The main aim of motor learning is to gain control over the system. Particularly in sports, movements are too complex to be acquired at once. Rather, movements are divided into parts (exercises) and these parts are exercised separately and in isolation before they are recombined to gradually build the whole skill. For example, to learn the high jump: run-up and take-off (note: run-up and take-off should not be isolated) as well as clearing the bar may be separately exercised before run-up, take-off and clearing the bar are recombined at a later stage of the learning process.

Acquiring motor skills can take place in different forms:

- Learners can engage in a trial-and-error (or rather trial-and-success) process consisting of a succession of (more or less deliberate) trials and evaluation of the outcomes resulting in an iterative cycle of planning, trying and evaluation.
- Learners can be guided by a teacher who may show the movement or exercise first. In the learning process, the teacher may use positive and negative reinforcement techniques to gradually shape the desired movement.

- Learners can initially engage in a cognitive solutions process. After having found the correct solution, they may be immediately able to perform a correct or nearly correct motor activity.

Of course, learners (and teachers) may also switch between different learning methods.

In the next section, examples for modelling motor learning in both humans and robots will be presented.

Concepts and methods of modelling

The problems described in the previous section require adequate modelling. First, models of skill structure will be introduced, followed by modelling motor control and motor learning, respectively.

Models of skill structure

Models of skill structure can be classified into qualitative and quantitative models (see Figure 1.3).

The simplest quantitative model of a motor activity is a single outcome measure. In many sports, a "centimeter–gram–second" (CGS) outcome can be determined, like jumping height, lifted weight and achieved time. In other sports, discrete states are determined, like goal or basket. Another important measure of motor activity used in exercise and health sciences is energy expenditure (EE), expressed either in terms of oxygen uptake or metabolic equivalents (METs, i.e., working EE related to resting EE).

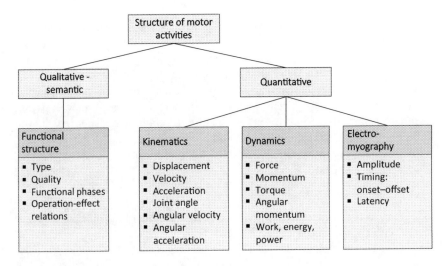

Figure 1.3 Overview and examples of qualitative and quantitative models of motor activities.

As has been pointed out in the previous section, motor activities can be structured into components. In sport science, several semantic models have been proposed to perform a functional analysis of motor activity. The concept proposed by Göhner (1979) distinguishes certain "functional phases" related to five reference conditions: movement goals, characteristics of the person, characteristics of the device to be moved, environment and rules. These functional phases can be structured according to temporal order (preparing, supporting or transitional phases) and functional significance or dependency (main phases or phases of first, second or third order). Preparing phases precede other phases, whereas supporting phases are performed simultaneously and transitional phases are performed successively. Main phases are the most important, functionally independent phases, whereas phases of first, second or third order depend on the respective superordinate phase, i.e., main, first- or second-order phase.

Using the Fuzzy logic approach, Schiebl (2003) analyzed different sport skills by representing the functional phases as if-then statements (e.g., for the lifting movement in rock and roll: If the weight of the female is low, and both height and ability indices are low, then the male should start lifting with an almost extended knee joint).

An example of Fuzzy sets is illustrated in Figure 1.4. In Figure 1.4a, three Fuzzy sets of the linguistic variable "height" comprising the terms "low, moderate, high" have been defined, based on the basic variable "height index male/female." In Figure 1.4b, three Fuzzy sets for the linguistic variable "height of the female" have been defined. As a characteristic of Fuzzy sets and in contrast to crisp logic, certain values can be elements of two or more Fuzzy sets at the same time. The simulation process in Fuzzy logic consists of three stages: fuzzification of the respective values, calculating the if-then rules and defuzzification. Therefore, continuous values are transformed into Fuzzy sets to operate on these sets, and finally Fuzzy sets are retransformed to continuous values.

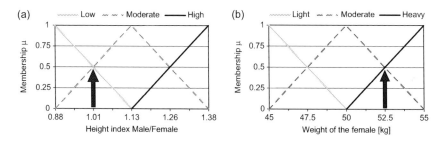

Figure 1.4 Examples of two Fuzzy sets for rock and roll. Abscissa: basic variable; ordinate: degree of membership. Three Fuzzy sets are illustrated: low/light, moderate and high/heavy, respectively. (A) A value of 1.01 (black arrow) is a member of the set "low index" (0.5) and "moderate index" (0.5); (B) a weight of 52.5 kg is a member of the set "moderate weight" (0.5) and "heavy weight" (0.5).

Another approach predominantly used in biomechanics applies physical or advanced statistical methods to structure movements. One option is to apply deterministic physical models to the task. Taking the example of throwing, a deterministic law exists for the prediction of throwing distance as a function of angle, velocity and height of release. For example, Müller and Sternad (2009) applied this function to aimed throwing to determine four factors or strategies contributing to performance variability (and stability). Another option to reduce the complexity of motor skills is to use principal components analysis (PCA). For example, Kollias, Hatzitaki, Papaiakovou and Giatsis (2001) applied PCA to reveal the biomechanical structure of vertical jumps.

Another option is to combine factor analysis and multiple regression. By applying factor analysis, the huge number of possible variables is reduced to a few factors. In a subsequent regression analysis, the (independent) contribution of the remaining factors (or variables) is determined. This approach has been applied, for example, by Schwameder and Müller (1995) to ski jumping. They identified 11 out of 63 biomechanical variables using a factor analysis. A regression analysis including these 11 variables revealed the significant influence of 4 variables (e.g., body-ski angle) on jumping width.

Classification models

Classifying motor skills requires assigning a specific motor activity to distinct classes according to certain assignment rules. Classification models can be simple or complex as well as implicit or explicit. Examples of simple models are the open-versus-closed skills dichotomy or the discrete-versus-cyclic dichotomy. Another simple classification model is grading, i.e., classifying movement quality according to a ranking scale (e.g., from "very good" to "very bad") or skill level (e.g., "novice," "advanced," and "expert"). More sophisticated two-dimensional models relate situational conditions (e.g., constant or variable) to the kind of motor response (e.g., constant or variable). The above-mentioned models are examples of top-down or explicit modelling; here, classification is derived from an explicit or white-box model.

In computer science in sport, black-box models are often used. When the number of classes is not known in advance, "classical" statistics can be used like cluster analysis. Another option is to use unconventional modelling like ANNs. ANNs are computational approaches taking selected characteristics of natural neural networks into consideration. ANNs can be subdivided into supervised and unsupervised learning (e.g., Perl, 2010). These approaches have gained much popularity. Therefore, it is not surprising that ANNs have been included in software packages like MATLAB® and SPSS®.

Particularly in throwing movements, ANNs have often been applied to HMA, in particular self-organizing maps (SOM) or Kohonen Feature Maps (KFM; e.g., Eimert, 1997; Bauer & Schöllhorn, 1997; Schmidt, 2012; Lamb, Bartlett & Robins, 2010). SOM and KFM are classes of unsupervised ANNs; this type of ANN classifies patterns or features according to similarity by self-organization,

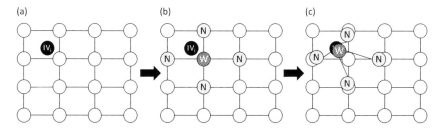

Figure 1.5 Illustration of a two-dimensional 4 x 4 KFM. (A) Input vector IV_i is presented to the network; (B) A winner neuron (W) is determined most closely representing the input vector; (C) Winner neuron (W) and neighbors (N) change their vectors toward the input vector.

i.e., similar patterns are spatially mapped to the same parts of the network (see Figure 1.5).

In Figure 1.5, an example of a 4 x 4 KFM is illustrated. After initialization, the first example (input vector IV_i; Figure 1.5a) is presented to the KFM. The neuron most closely representing the input vector is identified (winner neuron W, Figure 1.5b). Subsequently, the winner neuron changes its vector toward the input vector (Figure 1.5c). In a similar but reduced manner, the four neighbor neurons also change (Figure 1.5c). This process runs iteratively until a certain criteria is fulfilled, e.g., a predefined number of simulation runs.

Perl (2004) identifies two important disadvantages of KFM: the large amount of data required for training and the limited capability to continue learning or dynamically adapt to new patterns. As a solution, he proposed the Dynamically Controlled Network (DyCoN), which is able to learn with fewer examples and to dynamically adapt to new patterns.

Eimert (1997) applied KFM to the classification of shot put. She used a 10 x 10 KFM to classify 100 movement trials of 7 participants according to 68 biomechanical features. As a result, the KFM was able to identify individual techniques as well as different quality levels.

Bauer and Schöllhorn (1997) applied SOM to discuss throwing (2 athletes; 45 trials of a decathlete, 8 trials of a specialist). For each trial, 51 time-normalized 34-dimensional feature vectors were used. Using a 11 x 11 SOM, the trials of the decathlete could clearly be separated from the specialist. Furthermore, the different training and competition sessions could be identified using the distance values. In another study, Schöllhorn and Bauer (1998) were able to identify male and female athletes as well as individual styles using a three-dimensional 7 x 6 x 6 network. In this study, 51 (time-normalized) 34-dimensional feature vectors of 8 male and 19 female javelin throwers (total: 50 trials) were included.

Lamb et al. (2010) coupled two SOM architectures (phase SOM1: 42 x 13 neurons; trial SOM2: 9 x 6 neurons) for classifying three different throwing techniques (three-point shot, free-throw shot and hook shot) in basketball (sample: four throwers; number of trials: not reported). Based on biomechanical data of the throws (i.e., normalized 2D position data in the sagittal plane for the right and left ankles, knees, hips and shoulders), SOM1 classified the biomechanical data into different movement phases, e.g., preparation, extension and release phase, whereas SOM2 located the players within a 2D space.

Schmidt (2012) applied a two-dimensional 10 x 10 DyCoN to classify free-throwing movements in basketball (sample: 21 throwers of different skill level; 20 throwing trials for each player; total: 420 trials). Based on the biomechanical data (angular displacements and angular velocities of the foot, knee, hip, shoulder, elbow and hand joints of the throwing side) of 105 shots (five trials per participant), she was able to identify movement phases as well as experts and novices.

Computational models of motor control

Skilled human performance means that humans are able to realize solutions for a motor task fulfilling the multiple constraints. For several tasks, simple spatio-temporal constraints could be identified resulting in computational models of task space, i.e., the corridor, set or manifold of possible solutions. Representative examples are catching or hitting a ball approaching the human (e.g., Lee, 1980; Savelsbergh, Whiting, Pijpers & Van Santvoord, 1993; Caljouw, Van der Kamp & Savelsbergh, 2004), aimed throwing (e.g., Müller & Loosch, 1999; Müller & Sternad, 2009), and juggling (e.g., Beek, 1989). Catching moving objects with or without self-motion is an example for modelling perceptual constraints on action. To catch or hit an approaching ball, humans can rely on or control simple features like time to contact (Lee, 1980), vertical acceleration (Michaels & Oudejans, 1992) or bearing angle (Lenoir, Musch, Thiery & Savelsbergh, 2002). When throwing at a target, particular spatial corridors determine whether throws are successful or not. In order to achieve a particular throwing width or end position, an equation including height of release (in relation to height of target), angle of release and initial velocity can be applied to determine how far the thrown object will fly (see Figure 1.6).

Beek (1989) identified specific temporal and spatial constraints on juggling. For example, he could show that to avoid collision of the balls, certain spatial regions have to be considered. These regions depend on the number of balls and the throwing parameters.

The approach of internal models has been introduced into neuroscience and cognitive science as "neural mechanisms that can mimic the input/output characteristics, or their inverses, of the motor apparatus" (Kawato, 1999, p. 718). A simple model comprising an inverse and a forward model is illustrated in Figure 1.7. The inverse model controls the prediction of appropriate motor commands to establish a desired movement, and the forward model controls the prediction of sensory feedback and outcome based on a copy of the specified motor commands (efference copy).

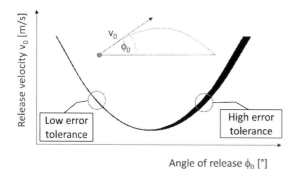

Figure 1.6 Arbitrary task space for an aimed throwing movement. The black area represents combinations of v_0 and φ_0 leading to a hit. Note that when φ_0 is increased beyond a certain value, the tolerance for changing v_0 increases; aerodynamic properties due to the shape of the object are neglected.

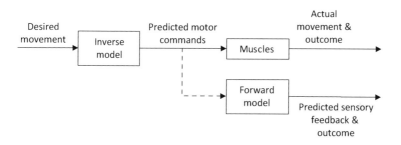

Figure 1.7 Internal models. The inverse model controls the prediction of motor commands to establish the desired movement, whereas the forward model predicts sensory feedback and outcome based on motor commands.

Computational models of motor learning

The idea to compare the human brain with a sophisticated machine and particularly a computer is very old (computer metaphor). Starting from naive and oversimplified comparisons, computational approaches to motor learning have become increasingly sophisticated and try to improve representation of neurophysiological as well as psychological mechanisms. In this regard, much progress has been made (reviews: Shadmehr, Smith & Krakauer, 2010; Wolpert & Flanagan, 2016; Dhawale, Smith & Ölveczky, 2017). Taking the example of internal models mentioned in the previous section, computational modelling could ask the question how the human sensori-motor system acquires these models. One example for an early attempt in this regard is the work by Künzell (1996). He examined the context interference effect when learning aimed throwing using a five-layer artificial neural network. Using backpropagation algorithms, he first trained the forward model to predict throwing width based on angle of release and release velocity. In a second step, the inverse model was trained to predict angle of release and release velocity depending on the desired throwing distance. Meanwhile, much more sophisticated computational approaches to learning have been developed, e.g., deep learning (reviews: Schmidhuber, 2015; LeCun, Bengio & Hinton, 2015).

> Deep-learning methods are representation-learning methods with multiple levels of representation, obtained by composing simple but non-linear modules that each transform the representation at one level (starting with the raw input) into a representation at a higher, slightly more abstract level.
>
> (LeCun et al., 2015, p. 436)

Therefore, low-dimensional and high-dimensional features of complex input can be detected and combined. Deep learning has been applied predominantly to cognitive tasks. One example of an application to motor learning has been presented by Berniker and Kording (2015). The authors used a deep neural network (i.e., stacked autoencoder) to approximate optional trajectory functions for motor control of reaching, depending on initial and desired final states. In a first step, they trained a deep network to find low-dimensional features characterizing optimal output functions. In a second step, a shallow supervised network was trained to map the input to these low-dimensional features of the output function.

A further example is the coupling of deep networks and reinforcement learning (RL). In RL, an agent learns to act in an environment guided by rewards and values in an unsupervised manner, i.e., by trial and error (introduction: Sutton & Barto, 1998). An RL system comprises four components: policy, reward, value and model of the environment. A policy is a mapping of environment and actions. A reward is the numerical representation of the (instantaneous) desirability of a state, whereas a value is the representation of a total reward that can be expected in the future, i.e., the long-term desirability. The model of the environment allows to predict future states and rewards based on current states and actions. When applying RL to deep networks, a common issue is lack of stability. Two possible solutions can be either introducing an experience replay memory or to have multiple agents act in parallel and asynchronously in multiple instances of environment (Mnih et al., 2016).

Furthermore, models of motor learning have been applied to robotics: "The easy analogies between the motor system and robotics have long fueled synergies between these fields" (Berniker & Kording, 2015, p. 1). An early example is the use of computational imitation models (e.g., Schaal, Ijspeert & Billard, 2003). As a more recent example, Mülling, Kober, Kroemer and Peters (2013) developed a framework that allows robots to learn cooperative table tennis with a human by mapping states to corresponding actions. Depending on specific state information, the system selects and generalizes motor primitives stored in a library. Applied to learning table tennis, the robot first learns a set of striking movements guided by the human ("kinesthetic teach-in") and extracts movement primitives. Second, the system learns and subsequently generalizes augmented states specifying predicted time, location and way of hitting the ball by using an RL algorithm. Therefore, the task of the system is to select and probably adapt the appropriate set of motor primitives according to the augmented state.

Methodological experience

All the models of motor activity introduced above have specific strengths and weaknesses. These will be discussed in the following section.

In general, as holds for any modelling procedure, there is the danger of taking the model for the original; models are always reductive representations of the complex original. Therefore, models always have restricted or limited validity constrained by more or less specific boundary conditions. Neglecting these constraints may lead to an overestimation of the model.

Semantic skill models on the one hand seem to be intuitive and straightforward. On the other hand, no clear rules for movement analysis are given or given criteria are lacking precision. Furthermore, no standard procedure is given. As a consequence, the modelling process and its result are rather arbitrary.

Computational classification models of skills are dominated by black-box models. An important step required to understand what is in the black box is to explore the classification behavior of the model. Unfortunately, this is often done very superficially. For example, good or bad trials as well as trials performed by distinct individuals are used for evaluation.

Mechanical models can be used to get a first impression of the constraints and tolerances that are relevant for the respective task. For many tasks, this analysis is a required, but not sufficient, initial step that has to be followed by subsequent analysis of real movements.

Internal models are a parsimonious approach explaining skilled behavior. These models are still very simple and intuitive. However, transfer and generalization problems arise as well as empirical validation. For the cerebellum, the approach of internal models seems to be well accepted. Another issue is how to implement this rather generic approach.

Results

Semantic models of skills are well accepted in many fields of sport practice. Determining the phase and functional structures of sport skills is a first step to methods of teaching. Depending on the particular model, different methods can be applied, e.g., from the simple to the complex, from the most important to the less important or from the start to the beginning. Taking the example of throwing, the skill can be divided into a preparatory or initial, a main and a final phase. The main phase can be further subdivided into different operations including movements of the hip, the trunk and the throwing arm. All these operations have to be performed in a particular spatio-temporal order to result in a successful throwing movement.

Mechanical models are also well accepted in research and practice. Taking the example of (aimed) throwing at a target, this approach has revealed specific spatio-temporal constraints and tolerances regarding the solution manifold (see Figure 1.6). In addition, specific strategies for reducing errors can be derived from the model (Müller & Sternad, 2009). On the one hand, one can shift performance into areas of high tolerance, i.e., from lower to higher angles of release. Another strategy is to decrease variability of performance in order to stay within the area of tolerance (noise reduction). A third strategy is to exploit and establish covariations, i.e., coupling angle of release and velocity of release as depicted in Figure 1.6.

Computational models of motor control have revealed numerous representational processes. Much work has concentrated on predictive and feedback control, i.e., internal forward and inverse models. For example, the role of the cerebellum in error processing has been investigated.

Computational models of motor learning have addressed numerous questions and aspects of memory, retention and transfer of motor skills, e.g., time course of motor learning (at least two distinct processes: one probably explicit fast process, and one probably implicit slow process), various kinds of interference and the significance of rewards in error-based (adaptive) versus structure learning (Shadmehr et al., 2010; Wolpert, Diedrichsen & Flanagan, 2011; Wolpert & Flanagan, 2016). Beyond various approaches of reinforcement and machine learning, ANNs have been successfully applied to contextual interference. Particularly in this application area of black-box modelling, transfer of the models to the human original is an issue. Unless the intrinsic structure of the model is thoroughly explored and functional and/or anatomical correspondence to the human sensori-motor system can be established, successful simulations just prove one effective solution in the space of solution manifold, but no valid "proof" for a solution actually implemented by the human system.

Examples and exercises

Vertical jump

The vertical jump was chosen as an example for the reader to develop own models of motor activities. The task is to jump as high as possible by performing a countermovement (CMJ) jump (Figure 1.8).

Motor activity: control and learning 17

Figure 1.8 Illustration of a countermovement jump (CMJ).

Task 1: Movement structure.
Look at Figure 1.8 and try to identify phases of the movement. Second, try to identify operations of particular body parts and their functions as well as the timing of these operations.

Task 2: Biomechanical model of jumping height.
Try to develop a biomechanical model of the CMJ to determine jumping height.

Hint: There are at least three possible approaches based on temporal or dynamic variables determining the achieved height.
How can the contribution of different body parts (legs, arms) be modeled?

Task 3: Develop a computer simulation for analyzing the CMJ.
Suppose that biomechanical data of a CMJ are available. Here, we assume that vertical forces are measured by a force platform (Figure 1.9). Your task is to develop a program that is able to analyze the biomechanical data of a CMJ. In particular, the vertical impulses and jumping height should be determined.

Aimed throwing

Task 4: Try to recalculate the areas of tolerance for aimed throwing at different targets, e.g., dartboard and basketball (Figure 1.6; Müller & Sternad, 2009).

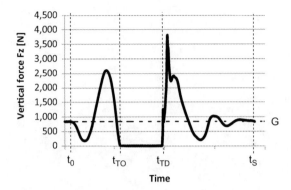

Figure 1.9 Vertical forces of a CMJ (own data); t_0 – movement onset; t_{TO} – take-off; t_{TD} – touchdown; t_S – quiet stand.

Conclusions

There is great variety in HMA. Therefore, numerous attempts have been undertaken to classify them. However, the respective classification system is always based on a particular reference model. Therefore, no generally agreed or unique classification system is available.

Modelling HMA can take several forms, ranging from the structure of movements over motor control to motor learning. HMA is an interdisciplinary phenomenon including biomechanical, (neuro-)physiological and psychological aspects. Therefore, many models of HMA can be attributed to these disciplines as well as their intersections.

Regarding types of computational models, all kinds of models have been applied, e.g., black-box and white-box models, structural and procedural models, etc. For example, ANNs have been applied for modelling motor learning and classifying or evaluating HMA. Currently, approaches of machine learning and cognitive science are gaining importance.

There are two main application purposes, i.e., explanation of the original "human motor activities (in sport)" and developing technical solutions, e.g., for robotic learning. Whereas the latter is not considered critical from an epistemological point of view due to the criteria of technical feasibility, the former may pose the issue of validity. In this regard, models claiming to explain HMA must verify functional and/or structural correspondence or at least plausibility with the original.

References

Argyropoulos, G. (2016). The cerebellum, internal models and prediction in 'non-motor' aspects of language: A critical review. *Brain and Language*, *161*, 4–17.

Bauer, H. & Schöllhorn, W. (1997). Self-organizing maps for the analysis of complex movement patterns. *Neural Processing Letters*, *5*(3), 193–199.

Beek, P. (1989). Timing and phase locking in cascade juggling. *Ecological Psychology, 1*(1), 55–96.
Beek, P., Peper, C. & Stegeman, D. (1995). Dynamical models of movement coordination. *Human Movement Science, 14*(4), 573–608.
Berniker, M. & Kording, K. (2015). Deep networks for motor control functions. *Frontiers in Computational Neuroscience, 9*.
Bernstein, N. (Ed.). (1967). *The coordination and regulation of movements.* Oxford: Pergamon Press.
Caljouw, S., Van der Kamp, J. & Savelsbergh, G. (2004). Catching optical information for the regulation of timing. *Experimental Brain Research, 155*(4), 427–438.
Dhawale, A., Smith, M. & Ölveczky, B. (2017). The role of variability in motor learning. *Annual Review of Neuroscience, 40*(1), 479–498.
Eimert, E. (1997). Beobachten und Klassifizieren von sportlichen Bewegungen mit neuronalen Netzen [Observing and classifying sport movements by means of neural networks]. Unpublished Ph.D. thesis, University of Tübingen.
Göhner, U. (1979). *Bewegungsanalyse im sport [Movement analysis in sport].* Schorndorf: Hofmann.
Kawato, M. (1999). Internal models for motor control and trajectory planning. *Current Opinion in Neurobiology, 9*(6), 718–727.
Kelso, J. (1997). Relative timing in brain and behavior: Some observations about the generalized motor program and self-organized coordination dynamics. *Human Movement Science, 16*(4), 453–460.
Kollias, I., Hatzitaki, V., Papaiakovou, G. & Giatsis, G. (2001). Using principal components analysis to identify individual differences in vertical jump performance. *Research Quarterly for Exercise and Sport, 72*(1), 63–67.
Künzell, S. (1996). *Motorik und Konnektionismus. Neuronale Netze als Modell interner Bewegungsrepräsentationen [Motor control and connectionism. Neural networks as model of internal representations].* Köln: bps.
Latash, M. (2010). Stages in learning motor synergies: A view based on the equilibrium-point hypothesis. *Human Movement Science, 29*(5), 642–654.
Lamb, P., Bartlett, R. & Robins, A. (2010). Self-organising maps: An objective method for clustering complex human movement. *International Journal of Computer Science in Sport, 9* (2), 20–29.
LeCun, Y., Bengio, Y. & Hinton, G. (2015). Deep learning. *Nature, 521* (7553), 436–444.
Lee, D. (1980). 16 Visuo-motor coordination in space-time. *Advances in Psychology, 1*, 281–295.
Lenoir, M., Musch, E., Thiery, E. & Savelsbergh, G. (2002). Rate of change of angular bearing as the relevant property in a horizontal interception task during locomotion. *Journal of Motor Behavior, 34*(4), 385–401.
Marr, D. (1982). *Vision: A computational investigation into the human representation and processing of visual information.* San Francisco: Freeman and Company.
Maurer, L., Maurer, H. & Müller, H. (2015). Neural correlates of error prediction in a complex motor task. *Frontiers in Behavioral Neuroscience, 9*, Article 209.
Michaels, C. & Oudejans, R. (1992). The optics and actions of catching fly balls: Zeroing out optical acceleration. *Ecological Psychology, 4*(4), 199–222.
Mnih, V., Badia, A., Mirza, M., Graves, A., Lillicrap, T., Harley, T., ... Kavukcuoglu, K. (2016, June). Asynchronous methods for deep reinforcement learning. In Proceedings of The 33rd International Conference on Machine Learning (pp. 1928–1937). New York, NY: PMLR.
Müller, H. & Loosch, E. (1999). Functional variability and an equifinal path of movement during targeted throwing. *Journal of Human Movement Studies, 36*, 103–126.

Müller, H. & Sternad, D. (2009). Motor learning: Changes in the structure of variability in a redundant task. In D. Sternad (Ed.), *Progress in motor control* (pp. 439–456). New York: Springer.

Mülling, K., Kober, J., Kroemer, O. & Peters, J. (2013). Learning to select and generalize striking movements in robot table tennis. *The International Journal of Robotics Research, 32*(3), 263–279.

Newell, K. & Vaillancourt, D. (2001). Dimensional change in motor learning. *Human Movement Science, 20*(4), 695–715.

Perl, J. (2004). A neural network approach to movement pattern analysis. *Human Movement Science, 23*(5), 605–620.

Perl, J. (2010). Net-based phase-analysis in motion processes. *Mathematical and Computer Modelling of Dynamical Systems, 16*(5), 465–475.

Savelsbergh, G., Whiting, H., Pijpers, J. & Van Santvoord, A. (1993). The visual guidance of catching. *Experimental Brain Research, 93* (1), 148–156.

Saziorski, W., Aruin, A. & Selujanow, W. (1984). *Biomechanik des menschlichen bewegungsapparates [Biomechanics of the human movement system]*. Berlin: Sportverlag.

Schaal, S., Ijspeert, A. & Billard, A. (2003). Computational approaches to motor learning by imitation. *Philosophical Transactions of the Royal Society B: Biological Sciences, 358*(1431), 537–547.

Schiebl, F. (2003). Fuzzy-bewegungsanalyse [Fuzzy movement analysis]. *Sportwissenschaft, 33,* 48–61.

Schmidhuber, J. (2015). Deep learning in neural networks: An overview. *Neural networks, 61,* 85–117.

Schmidt, A. (2012). Movement pattern recognition in basketball free-throw shooting. *Human Movement Science, 31*(2), 360–382.

Schmidt, R. (1980). On the theoretical status of time in motor program representations. *Advances in Psychology, 1,* 145–166.

Schmidt, R. (2003). Motor schema theory after 27 years: Reflections and implications for a new theory. *Research Quarterly for Exercise and Sport, 74*(4), 366–375.

Schneider, K., Zernicke, R., Schmidt, R. & Hart, T. (1989). Changes in limb dynamics during the practice of rapid arm movements. *Journal of Biomechanics, 22*(8–9), 805–817.

Schöllhorn, W. & Bauer, H-U. (1998). Identifying individual movement styles in high performance sports by means of self-organizing Kohonen maps. In H. Riehle & M. Vieten (Eds.), Proceedings of The XVI ISBS 1998. (pp. 574–577). Konstanz: Konstanz University Press.

Schwameder, H. & Müller, E. (1995). Biomechanische beschreibung und analyse der V-technik im skispringen [Biomechanical description and analysis of the V technique in ski jumping]. *Spectrum der Sportwissenschaft, 7*(1), 5–36.

Shadmehr, R., Smith, M. & Krakauer, J. (2010). Error correction, sensory prediction, and adaptation in motor control. *Annual Review of Neuroscience, 33,* 89–108.

Summers, J. & Anson, J. (2009). Current status of the motor program: Revisited. *Human Movement Science, 28*(5), 566–577.

Sutton, R. & Barto, A. (1998). *Reinforcement learning: An introduction* (Vol. *1*, No. 1). Cambridge, MA: MIT Press.

Vereijken, B., Emmerik, R., Whiting, H. & Newell, K. (1992). Free(z)ing degrees of freedom in skill acquisition. *Journal of Motor Behavior, 24*(1), 133–142.

Whiting, H. (1984). *Human motor actions. Bernstein reassessed*. Amsterdam: Elsevier.

Wiemeyer, J. (2003). Function as constitutive feature of movements in sport. *International Journal of Computer Science in Sport, 2*(2), 113–115.

Wolpert, D., Diedrichsen, J. & Flanagan, J. (2011). Principles of sensorimotor learning. *Nature Reviews. Neuroscience*, *12*(12), 739–751.
Wolpert, D. & Flanagan, J. (2016). Computations underlying sensorimotor learning. *Current Opinion in Neurobiology*, *37*, 7–11.
Wolpert, D. Miall, R. & Kawato, M. (1998). Internal models in the cerebellum. *Trends in Cognitive Sciences*, *2*(9), 338–347.

2 Muscle mechanics

Sigrid Thaller and Harald Penasso

Introduction

Every movement is caused by forces. In human movements, these forces are either external forces (e.g., gravity) or forces due to muscle contractions. Therefore, muscle mechanics play an important role in models of human movements.

Mechanical modelling of muscles can take place at different hierarchical levels: at the sarcomere level (Schappacher-Tilp, Leonard, Desch & Herzog, 2015), describing for example the actin-myosin-titin interaction; at the fiber level (MacIntosh, Herzog, Suter, Wiley & Sokolosky, 1993), including different fiber types; at the whole-muscle level (Hill, 1970); or at the muscle-group level (Bobbert, 2012). The force-velocity dependence of the muscle gives information on, among other things, the efficiency and the fiber distribution. In addition, the elastic properties of tendons and other elastic structures influence energy storage and the behavior of the exerted force over time.

Important issues in modelling human movements are the interaction of the muscle with its activation and its geometrical conditions, and how the exerted force of the neuromuscular system is related to the accelerating force in a movement.

In the next section, we state some goals of biomechanical modelling. *Concepts and methods for modelling* describes important concepts for modelling that are necessary for reaching these goals. In *Methodological experience*, we describe the physiological and mechanical background of muscle properties in order to develop appropriate relations on muscle mechanics. In *Results*, we sketch a model for a simple movement that can be used for simulations as well as for finding individual muscle parameters. Sensitivity and dimension analysis of models lead to further results useful for sports science. We conclude this chapter on muscle mechanics with some applications and exercises.

Project/problem

There are numerous models of human movements in literature ranging from very complex (e.g., Hatze, 1976, 1977) to very simple ones (e.g., Bobbert, 2012). The degree of complexity depends on the specific goal of the model (e.g., whether the model is used for description or prediction of a movement, or the level of the resulting accuracy). Usually, a model of a human movement is a system of nonlinear differential equations. We want the parameters in the equations to be constants with a clear physiological or physical meaning. Either they should be conditions of the movement (e.g., gravitation, temperature, air pressure, mass of sports equipment) or properties of the subject (e.g., mass, leg length, isometric force of a fully activated muscle). In models that are used for predictions, e.g., for the simulation of the effect of changes in the conditions on the movement, the movement independence of the parameter values is crucial. However, parameters that describe a mixture of conditions and properties (e.g., rate of force development) depend on the particular movement and thus should be avoided (see also the next section on force laws).

This chapter deals with three goals of modelling muscle mechanics: First, we want to find models using only movement-independent parameters. For predicting individual differences, we have to know the individual input parameter values. Thus, in a second step, we look for models of specific movements that allow us to calculate the individual parameter values by comparing the model output to measurements of the movements. The third problem is to find individual differences by looking at the sensitivity and at dimensionless parameters of the model.

Concepts and methods for modelling

Movement-independent force laws

By Newton's law, the sum of the acting forces F on a mass m causes the acceleration a of the mass m, $\sum F = m \cdot a$. If the initial state, the mass and the acting forces are known, this equation of motion (a second-order differential equation) yields the position of the mass as a function of time.

For example, neglecting any air drag, the knowledge of the mass m and the ground reaction force $F(t)$ determines the position of the center of mass (COM) in a squat jump at every instant of time. The acceleration $a(t)$ of the COM as a function of time is given by $a(t) = F(t)/m - g$, where g is the gravitational acceleration. By integrating, one gets the velocity $v(t)$ as $v(t) = v(0) + \frac{1}{m}\int (F(t) - g \cdot m)dt$, $v(0)$ denoting the measured initial velocity. A further integration leads to the COM position as a function of time.

The disadvantage of this approach is that the measured force $F(t)$ depends on the measured movement and thus no predictions can be made about other movements, for example with different initial positions or with different masses. This kind of model is called an empirical or phenomenological model. For many purposes, empirical models are sufficient, for example if one wants to know net torques in the joints for a specific movement.

Let us consider the simple example of a linear elastic spring. We fix one end of the spring on the ceiling and put a load on the other side such that the spring will be stretched to an equilibrium position. Then we extend the spring by pulling the load a little bit before releasing it. The spring will oscillate around the equilibrium position. If we alter the amount of the load, the frequency of the oscillation will be different. In an empirical model, we would just determine the force as a function of time. This alone would not help us to answer questions about, e.g., the dependence of the frequency on the load.

However, we know that the force of an ideal elastic linear spring depends only on the displacement x. If we describe the elastic property of the spring movement independently by a constant k, $f = -k\,x$, then all these movements can be predicted without measuring just by evaluating the equation of motion $\ddot{x} = \dfrac{-k}{m} \cdot x$ including the initial conditions. The equation $f = -k\,x$ is called Hooke's law and is independent of the movement under consideration. The parameter k is the stiffness property of the spring and does not change during the movement. It can be determined by a single measurement. If a force is described by an equation using movement-independent parameters as in Hooke's law, this equation is called a force law. Combining such a force law with Newton's law leads to an equation of motion with parameters independent of the movement. This kind of model is called a structural or explanatory model; it allows predictions about the behavior of the system for different loads or different initial positions.

Direct and inverse dynamics

Models can be applied in two different ways: simulating a movement by evaluating the model equations using a set of fixed parameter values as input parameters (direct dynamics), or measuring a movement and finding a set of parameter values that fits to the measured movement (inverse dynamics).

For simulating a person's movement, the knowledge of all individual parameter values is necessary. In general, this is not possible. The activation of a muscle cannot be predicted completely. Only in special cases, e.g., when a muscle is activated with maximum voluntary effort, or in some cyclic movements, the activation as function of time, A(t), can be described using movement-independent parameters. Furthermore, not every single muscle and tendon property can be measured directly in vivo because data on, e.g., the force of each muscle and tendon, would be needed. However, such properties are required in complex models that are widely used, e.g., in gait analysis (e.g., Anderson & Pandy, 2001). Therefore, often average values are used, e.g., the shape of the force-velocity relation of a muscle, the stiffness of a tendon, the isometric force scaled to the cross-sectional area of a muscle and the mass of a body segment scaled to the total mass (Zajac, 1989; Hoy, Zajac & Gordon, 1990; Van Soest, Schwab, Bobbert & Van Ingen Schenau, 1993; Van Soest & Bobbert, 1993; Winters, 1995; Thelen, 2003; Winby, Lloyd & Kirk, 2008; Bayer, Schmitt, Günther & Haeufle, 2017).

When models are used for describing a measured movement, usually scaled average values of contractile and elastic properties are taken from literature and the activation is optimized to fit the measurements. As this method neglects many inter-individual muscular differences, one has to be careful when interpreting the results. In most cases, such models are underdetermined, meaning that different sets of parameter values lead to identical solutions (Crowninshield & Brand, 1981; Nigg & Herzog, 1999, p. 533). Imagine watching a competition in synchronized swimming. Although the movements are alike, we know that each athlete has individual muscle properties. Thus, it is not possible to find the individual parameters just by measuring these movements.

However, if the activation of every muscle in a simple movement is known and an appropriate simple model can be formulated to describe this movement, muscle parameters can be identified using inverse dynamics. As the parameters identified with this method are independent of the movement, they may be used as input parameters for other, more complex models.

In *Results*, we present a model that is appropriate for the determination of muscle parameters using inverse dynamics.

Sensitivity analysis

A model should always be as simple as possible and as complex as necessary for the modelling goal. In some cases, very sophisticated models are needed; in other cases, very simplified models are eligible. However, all parameters in the model should influence the model output. A sensitivity analysis of a model tests how much parameter value changes affect the solution of the model equations (Rockenfeller, Günther, Schmitt & Götz, 2015; Scovil & Ronsky, 2006). If the model output is not very sensitive to a parameter, this is a hint that this parameter may not be necessary for describing the system with regard to certain questions.

In *Results*, we show that sensitivity analysis gives some hints on the individual differences in the effect of training.

Methodological experience

Our goal is to describe muscle forces as movement-independent force laws. Therefore, let us have a very short look at how muscles work. For more details, we refer to books about muscle physiology (e.g., MacIntosh, Gardiner & McComas, 2016).

Sliding filament theory and cross-bridge theory

A skeletal muscle consists of muscle fascicles, which in turn consist of muscle fibers. Each fiber contains parallel myofibrils that are composed of sarcomeres in series. The sarcomeres are the basic functional units to produce force (Figure 2.1). They are approximately 2.64 µm long (Burkholder & Lieber, 2001) and show a striated pattern. They are separated from each other by the so-called Z-line. The sarcomere consists of aligned proteins. The thicker myosin filaments lie in the middle of the sarcomere; six thinner actin filaments surround each of them.

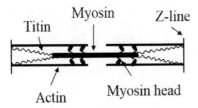

Figure 2.1 Schematic representation of a sarcomere. Actin filaments are connected to the Z-lines, the myosin heads bind the myosin filament to the actin filament and titin elastically connects myosin, actin and the Z-line.

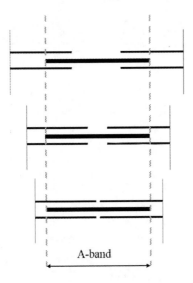

Figure 2.2 Sliding of the actin and myosin filaments against each other leads to a different overlap at lengthening, at optimal length and at shortening. The width of the A-band remains constant.

The actin filaments are connected to the Z-line between the sarcomeres. Another protein, titin, acts as an elastic connection between myosin, actin and the Z-line.

Our modern understanding of muscle contraction and how sarcomeres produce force is based on the works of H. Huxley and A. Huxley who independently published their theories in the same year (and at the same date in the same journal) (Huxley & Niedergerke, 1954; Huxley & Hanson, 1954). Using innovative microscopy techniques, they observed that the middle striation of the sarcomere, the A-band with the myosin filaments, does not change its length during a contraction. Therefore, they proposed that during a muscle contraction the actin and myosin filaments do not shorten but slide against each other (Figure 2.2). The force that lets the actin and myosin filament slide stems from the binding of the myosin heads to the actin, building the so-called cross-bridges (Huxley, 1957). During one cross-bridge cycle, the shape of the myosin necks changes and thereby it pulls the myosin against the actin. After releasing, the next binding process starts. The energy comes from adenosine triphosphate (ATP). Each cycle requires one ATP molecule. The sum of the forces produced by each cross-bridge is the force of the sarcomere. One can compare this to the forces at rope pulling or tug of war, where the individual forces sum up.

Length dependence of the muscle force and force enhancement

If we look at the length of the sarcomere, we see that the possible number of cross-bridges depends on the overlap of the myosin and actin filaments. In a contracted sarcomere, more bindings are possible at first, but at a certain level of overlap the structure itself will hinder further contraction (Figure 2.2). Rode, Siebert, Tomalka and Blickhan (2016) proposed that the myosin filaments even might slide through the Z-line. If the sarcomere is stretched, the overlap is smaller and thus fewer cross-bridges are acting. Therefore, there is an optimal length l_0 with maximal force (Gordon, Huxley & Julian, 1966). One has to be careful when using this relation between force and length as a functional dependence; always keep in mind that this relation was found by pointwise measurement, i.e., the force was measured isometrically at different lengths in different experiments and not during one movement. For a detailed discussion see, e.g., Epstein and Herzog (1998). There are numerous formulas acting as empirical models. For example, Van Soest and Bobbert (1993) use a parabola with an empirical constant 0.56, which determines the width of the parabola

$$f(l) = f_{iso}\left[1 - \left(\frac{l - l_0}{0.56 \cdot l_0}\right)^2\right], \tag{1}$$

where l denotes the length of the muscle, and l_0 the optimal length where muscle force is maximal. Epstein and Herzog (1998, p. 147) suggested the quadratic function

$$f\left(\frac{l}{l_0}\right) = -0.772 \cdot \left(\frac{l}{l_0}\right)^2 + 1.544 \cdot \left(\frac{l}{l_0}\right) - 0.494 \text{ for } 0.4 < \left(\frac{l}{l_0}\right) < 1.6 \tag{2}$$

to describe the stress (given in N/mm²) in fibers, and Woittiez, Huijing, Boom and Rozendal (1984) set

$$\frac{f(l)}{f(l_0)} = -6.25 \cdot \left(\frac{l}{l_0}\right)^2 + 12.5 \cdot \left(\frac{l}{l_0}\right) - 5.25 \text{ for } 0.6 < \left(\frac{l}{l_0}\right) < 1.4 \tag{3}$$

for the whole muscle (Figure 2.3).

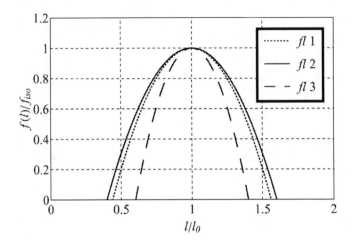

Figure 2.3 The three different models of the force-length relation in the muscle, *fl* 1, *fl* 2 and *fl* 3 (equations (1), (2) and (3) respectively) show the isometric muscle force $f(l)$ normalized to the isometric force at optimal length, $f(l_0)$.

Note that the constants given in (1)–(3) do not reflect any individual differences nor do they have any physical or physiological interpretation. Regardless of the fact that they are derived from isometric measurements, in many muscle models (e.g., Haeufle, Günther, Bayer & Schmitt, 2014; Hoy et al., 1990; Pandy, Zajac, Sim & Levine, 1990; Rajagopal, Dembia, DeMers, Delp, Hicks & Delp, 2016; Thelen, 2003; Van Soest & Bobbert, 1993; Yamaguchi & Zajac, 1989) equations like (1)–(3) are used as a multiplicative factor to muscle force even in concentric and eccentric movements. In these models, they are assumed to be movement independent and can therefore be used in force laws.

For some special movements, the length dependence of the muscle force may even be neglected because the muscles act in a length range where the change of force due to length changes is comparatively small (e.g., Hill, 1938).

Another very interesting topic related to the length dependence of the muscle force is the phenomenon of force enhancement and force depression (Herzog & Leonard, 2002). When activating a muscle at a given length, the muscle produces some isometric force. However, activating the muscle at a different length and then restoring the initial length leads to more (when elongated) or less (when shortened) isometric force. That means that there is some history dependence of the force production of the sarcomere. Early theories suspected that this could be caused by elastic structures in series. At isometric contraction, the sarcomere is not quite at rest but shortens while the elastic parts elongate. Changing the overall length now changes the length of the sarcomere in a non-proportional way. It may happen that the sarcomeres arranged in series in the myofibril obtain different lengths (sarcomere length non-uniformity) and this could cause the force enhancement. However, as the force enhancement could even exceed the force at the plateau region of the force-length relation (Leonard, DuVall & Herzog, 2010; Rassier, Herzog & Pollack, 2003), there must be another explanation for this phenomenon. A new hypothesis by Herzog (2014) takes the role of titin into account. Titin acts as a molecular spring that is calcium dependent; it binds to actin and thus influences the elastic properties. Furthermore, it may unfold its structure step by step and, therefore, change its length and elastic properties. This leads to a history-dependent behavior. A mathematical model of this situation can be found in Schappacher-Tilp et al. (2015). Heidlauf, Klotz, Rode, Siebert and Röhrle (2017) propose a different approach for modelling the titin-actin interactions.

Velocity dependence of the muscle force

Although the heat production and the force-velocity relation of muscles was already investigated earlier (Fenn, 1923; Fenn & Marsh, 1935; Gasser & Hill, 1924), the most cited paper on this topic is "The heat of shortening and the dynamic constants of muscle" by A. Hill (1886–1977) in which he investigated the heat production of isolated muscles (Hill, 1938). The energy E of an electrically stimulated muscle moving a constant load L is given by

$$E = A = ax + Lx, \tag{4}$$

where A denotes the constant activation heat, a is a constant heat coefficient with dimension of a force, L is the moved load and x is the change of position during the contraction. Hill neglected the kinetic energy because it is small in comparison to the mechanical work Lx and the energy loss due to friction ax. The experiments performed by Hill showed that the power output P of the muscle is proportional to the difference between maximal moveable load L_0 and moved load L

$$P = b \cdot (L_0 - L), \tag{5}$$

where b denotes the proportionality constant. As the time derivative of the energy E is the power output P, and the derivative of x is the velocity v, we get

$$b \cdot (L_0 - L) = (L + a) \cdot v.$$

This leads to Hill's equation

$$b \cdot (L_0 + a) = (L + a) \cdot (v + b). \tag{6}$$

As the original experiments were conducted isotonically, the moved load L was often used synonymously with the muscle force, thus giving a relation between muscle force and velocity (Hill, 1970). Substituting f instead of L and rewriting Equation (6) yields

$$f = \frac{b \cdot (f_{iso} + a)}{v + b} - a. \tag{7}$$

In the literature since 1938, the meaning of the quantities L and v in Equation (6) varies: For example, L sometimes denotes the load, sometimes the average force during the movement or the force as a function of the contraction velocity (Hill, 1938, 1970); v denotes the steady state velocity during the contraction (Hill, 1938), the mean velocity in an interval (Hill, 1970, p. 28) or the velocity at a certain position (Hill, 1970, p. 33). The substitution of the load L by the contraction force f in Equation (7) is not trivial because, originally, Hill found the relation between moved load and corresponding velocity by pointwise measurements in several experiments. One also has to be very careful in interpreting the meaning of the other involved quantities. For example, in the total extra heat $a \cdot x$, x denotes the change of position over the whole movement and only for a fixed velocity. That implies that x is not the position as a function of time. $a \cdot x$ does not denote the amount of heat energy that is released in the interval $[0, x]$ for every kind of movement but only for a movement with fixed load and corresponding velocity (see also Podolsky, 1961).

The sliding filament theory and the idea of cross-bridges led to more detailed models of muscle force. A. Huxley introduced rate functions f and g for the attachments and detachments of the myosin heads to the actin filaments. Roughly speaking, the probability of an attachment depends on the distance x between the equilibrium point of oscillating myosin heads and the attachment side on the actin filament and is given by $f(x)$. In an analogous way, the rate for detachment, $g(x)$, was defined.

Denoting as $n(t)$ the number of attached cross-bridges, he calculated the change of $n(t)$ via

$$\frac{dn(t)}{dt} = (1-n) \cdot f(x) - n \cdot g(x). \tag{8}$$

For a detailed discussion on the further implications of Huxley's equation (8) we refer in the literature to Herzog (2000), Huxley (1957) and Podolsky, Nolan and Zaveler (1969). Although this equation (8) and the derived force-velocity relation fit Hill's experiments very well (Huxley, 1957), there are some discrepancies that could be partly ascribed to the pointwise derivation of both Hill's and Huxley's equations (Hill, 1970; Huxley & Simmons, 1971). Huxley's formulation reveals more of the mechanism of muscle contraction but is mathematically far more complicated than Hill's equation. Therefore, in many muscle models, a formulation according to Hill is preferred.

The muscle force as a function of shortening velocity

Nowadays, Hill's equation

$$f(v) = \frac{c}{v+b} - a \tag{9}$$

with positive constants a, b and c, $c = b \cdot (f_{iso} + a)$, f_{iso} isometric force, is widely used to describe the functional dependence of the muscle force f on the shortening velocity v. This means that, during a movement, the force at any instant of time, $f(t)$, corresponds exactly to the velocity at this time, $v(t)$, according to Equation (9). As mentioned before, regarding Equation (7) as functional dependence, Equation (9) is not trivial at all. Sust (1978, 2009) pointed out that only under certain conditions (e.g., the acceleration of the load must be very small) the original relation between velocity and load fits the functional dependence between force and velocity in the muscle in (9) (see also Siebert, Sust, Thaller, Tilp & Wagner, 2007). Depending on the kind of measurements (isotonic, concentric, quick release …) and on the exact meaning of velocity (maximum velocity during the movement, velocity at the end of the measurement, average velocity …), some experiments seem to contradict (9). These difficulties can be overcome by using (9) in appropriate muscle models for describing the experiments. For example, even a linear dependence of force and velocity, both measured externally in some sports (Samozino, Morin, Hintzy & Belli, 2010), can be explained by applying (9) (Bobbert, 2012).

The positive parameters a, b and c in Equation (9) are movement-independent parameters. The parameter a is connected to friction and has the dimension of a force; the parameter b has the dimension of a velocity and is derived from the relation between force compared to maximum force and the power output; and c describes the chemical power. From Hill's Equation (9), we see immediately that the force has a maximum at velocity $v = 0$. This is the isometric force

$$f_{iso} = c/b - a. \qquad (10)$$

The largest contraction velocity v_{max} occurs when the force is 0,

$$v_{max} = c/a - b. \qquad (11)$$

Hill's equation is defined on the interval $[0, v_{max}]$, i.e., for isometric and concentric contractions. Figure 2.4 shows the graph of Hill's equation. It is a hyperbola with asymptotes at $v = -b$ and $f = -a$. As f_{iso} as well as v_{max} have to be positive, we get

$$ab < c. \qquad (12)$$

In Figure 2.4, ab is represented by the area of the rectangle between origin and asymptotes and c by the rectangle between the point $(0, f_{iso})$ and the asymptotes.

The mechanical power p is given by $p = fv$. As a function of the velocity, we can write the power as

$$p(v) = \left(\frac{c}{v + b} - a \right) \cdot v. \qquad (13)$$

The power is zero at $v = 0$ as well as at $v = v_{max}$, and it is positive in the open interval $(0, v_{max})$. In order to evaluate the optimum velocity v_{opt} at which the muscle can exert the maximum mechanical power, the derivative $\frac{dp(v)}{dv}$ has to be set to zero. This leads to

$$v_{opt} = \sqrt{\frac{bc}{a}} - b. \tag{14}$$

The optimum force f_{opt} at which the muscle produces the maximum power is the force at optimum velocity, $f(v_{opt})$, and it is given by inserting v_{opt} into Equation (9)

$$f_{opt} = \sqrt{\frac{ac}{b}} - a. \tag{15}$$

The maximum power p_{max} is now given as the product of f_{opt} and v_{opt},

$$p_{max} = f_{opt} \cdot v_{opt} = ab + c - 2\sqrt{abc}. \tag{16}$$

In Figure 2.4, the maximum power p_{max} is the largest rectangle that fits between the curve and the coordinate axes.

As Hill's equation describes a hyperbola in v-f-space, three independent parameters are necessary to describe this function. Using the constants a, b and c we get Equation (9), whereas using a, b and f_{iso} leads to a representation similar to Equation (7). In sports science, the use of f_{iso}, v_{max} and p_{max} is much more common than the use of a, b and c because these quantities are easier to interpret. f_{iso}, v_{max}

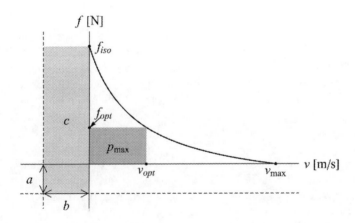

Figure 2.4 Hyperbolic shape of the force-velocity relation with asymptotes (dashed lines). In Hill's Equation (9), a and b are parameters and define the distances of the asymptotes from the axes. The areas of the rectangles represent $c = b(f_{iso} + a)$, ab, and maximum power $p_{max} = f_{opt} v_{opt}$. Thereby, fiso denotes the maximum isometric force, and f_{opt} and v_{opt} denote the optimum force and velocity.

and p_{max} can be expressed by the parameter set a, b and c, equations (10), (11) and (16). The conversion of a, b and c into f_{iso}, v_{max} and p_{max} needs some longer and tedious calculations; the result is given by

$$a = \frac{p_{max}}{v_{max}\left(1 - 2\sqrt{p_{max}/(f_{iso} \cdot v_{max})}\right)}, \tag{17}$$

$$b = \frac{p_{max}}{f_{iso}\left(1 - 2\sqrt{p_{max}/(f_{iso} \cdot v_{max})}\right)}, \text{ and} \tag{18}$$

$$c = \frac{p_{max}}{\left(1 - 2\sqrt{p_{max}/(f_{iso} \cdot v_{max})}\right)} + \frac{p_{max} \cdot p_{max}}{f_{iso} \cdot v_{max}\left(1 - 2\sqrt{p_{max}/(f_{iso} \cdot v_{max})}\right)^2}. \tag{19}$$

One can also use the optimum force or the optimum velocity as parameters in the representation of Hill's equation. For numerical computation in muscle models, some of these parameter sets are more convenient, depending on the calculations.

Eccentric force

Hill's Equation (9) is only valid for concentric and isometric movements. The eccentric force, i.e., the force developed for negative contraction velocities, is greater than the isometric and concentric force. In literature, there are several versions for extending Hill's isometric force-velocity relation to negative velocities. For describing the velocity dependence, at least one more parameter is needed. One possibility is to use another hyperbola with asymptotes $f_{ecc} = f_{ecc,max}$ and $v = B_{ecc}$, where $f_{ecc,max}$ denotes the maximum possible eccentric force, and B_{ecc} is some constant positive velocity. Again, for describing this hyperbola, i.e., the dependence of the eccentric force f_{ecc} on the velocity v, we need three quantities, $f_{ecc,max}$, B_{ecc} and some constant C_{ecc}.

$$f_{ecc} = \frac{C_{ecc}}{v - B_{ecc}} + f_{ecc,max}. \tag{20}$$

We impose the following conditions on the hyperbolic eccentric force-velocity relation: First, the force at zero velocity has to be f_{iso}, i.e., the eccentric and isometric curves are connected, and second, the slope of the eccentric and concentric curves at velocity zero is the same, i.e., the whole curve is differentiable at this point (Figure 2.5). The first condition leads to

$$f_{iso} = \frac{c}{b} - a = \frac{C_{ecc}}{-B_{ecc}} + f_{ecc,max}. \tag{21}$$

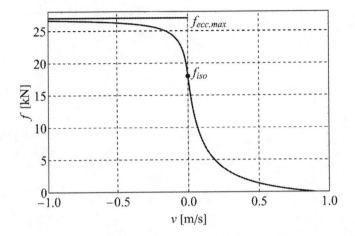

Figure 2.5 Eccentric ($v < 0$) and concentric ($v > 0$) force-velocity relation. The maximal eccentric force $f_{ecc,max}$ is given by $f_{ecc,max} = 1.5\, f_{iso}$ (cf. Equation (23)).

From the second condition, we get

$$\frac{-c}{b^2} = \frac{-C_{ecc}}{B_{ecc}^2} \qquad (22)$$

Combining the two equations (21) and (22), we get for $v \leq 0$ the eccentric force

$$f_{ecc} = \frac{\frac{b^2}{c}(f_{ecc,max} - f_{iso})^2}{v - \frac{b^2}{c}(f_{ecc,max} - f_{iso})} + f_{ecc,max} \qquad (23)$$

Instead of the second condition, there is also another assumption used in literature: The derivative df_{ecc}/dv of the eccentric curve at the point $v = 0$ is twice the derivative of the concentric curve at that point (Katz, 1939). In any case, there is only one parameter left for characterizing the eccentric extension and we can use, for example, $f_{ecc,max} = 1.5\, f_{iso}$ (Van Soest & Bobbert, 1993), or $f_{ecc,max} = 2\, f_{iso}$ (Simmons, 1992).

Using thermodynamic considerations and Huxley's cross-bridge theory, Mitsui and Ohshima (1988) developed a model of describing the eccentric and concentric force-velocity relation of the muscle. In a refined version (Stockhecker, 1995), the relation is given by

$$v = v_{max} \cdot \frac{\exp\left(A \cdot \left(1 - \frac{f}{f_{iso}}\right)\right) - \exp\left(-B \cdot \left(1 - \frac{f}{f_{iso}}\right)\right)}{e^A - e^{-B}}, \qquad (24)$$

where A and B are non-negative constants. Again, we have four quantities for describing the force-velocity relation for negative and positive contraction velocities. For the concentric case, Hill's relation (9) can approximate Equation (24). The second term in the denominator is related to the probability of back steps of the myosin heads in the bindings. If we consider this term to be constant, i.e., $B = 0$, the connection to Hill's relation (9) is approximately given by

$$a = \frac{f_{iso}}{A + A^2/2}, \; b = \frac{v_{max}}{e^A - 1}, \; c = \frac{f_{iso}}{A + A^2/2} \cdot \frac{v_{max}}{e^A - 1} \cdot e^A. \tag{25}$$

Hill's equation on different structural levels

It is an interesting fact that Hill's equation is not only valid for a single sarcomere but also for muscle fibers, muscle bundles and even for whole muscles. If we think of the structures put together by contractile units in series and in parallel then the question arises how a hyperbolic equation can fit to the smallest entity as well as to the whole unit.

Let us first consider a parallel arrangement of two contractile elements. There, the forces of the elements sum up, whereas the velocities have to be equal. Having two equations describing the forces of element 1 and element 2

$$f_1 = \frac{c_1}{v_1 + b_1} - a_1 \text{ and } f_2 = \frac{c_2}{v_2 + b_2} - a_2,$$

a parallel arrangement has to fulfill

$$f_p = f_1 + f_2 \text{ and } v_p = v_1 = v_2.$$

If the parameters b_1 and b_2 are equal, $b_1 = b_2$, we get

$$f_p = \frac{c_p}{v_p + b_p} - a_p \text{ with } a_p = a_1 + a_2, \; b_p = b_1 = b_2 \text{ and } c_p = c_1 + c_2. \tag{26}$$

This is a function of the same structure as in Equation (9).

In a serial arrangement of two elements, the forces are identical whereas the velocities sum up. Thus, we have $f_s = f_1 = f_2$ and $v_s = v_1 + v_2$. Combining two elements in series, now with equal parameters $a_1 = a_2 = a$, yields an analogous equation to (26):

$$f_s = \frac{c_s}{v_s + b_s} - a_s \text{ with } a_s = a_1 = a_2, \; b_s = b_1 + b_2 \text{ and } c_s = c_1 + c_2. \tag{27}$$

Assume the single elements are of different fiber types. The constant b was defined as a factor in Equation (5) and is related to the speed of chemical reactions

(Podolsky, 1960). Thus, one cannot expect that both elements have the same value for the parameter b. In order to get a hyperbolic force-velocity relation for the force $f_p = f_1 + f_2$ of parallel fibers, we may apply a Taylor expansion (see, e.g., Sust, Schmalz & Linnenbecker, 1997b):

$$f_p = \frac{c_p}{v_p + b_p} - a_p + R, \qquad (28)$$

where, again, $a_p = a_1 + a_2$, $c_p = c_1 + c_2$ and $b_p = (b_1 + b_2)/2$, and

$$R < \frac{2c_p}{b_2(r+1)} \cdot \left[\left(\frac{r-1}{r+1}\right)^2 + \left(\frac{r-1}{r+1}\right)^4 + \cdots \right], r = b_1 / b_2. \qquad (29)$$

For small differences between b_1 and b_2, the remainder R of the Taylor expansion can be neglected. A difference of 30% between b_1 and b_2 yields a 2.6% deviation of force in Hill's relation (9) (Siebert et al., 2007; Sust, 1987, p. 41).

The measurements that led to the derivation of Hill's equation also contained effects of elastic structures. However, the parameters should not contain mixed information on contractile and elastic structures because that could harm the movement independence of the parameters. Thus, models usually characterize the length and velocity dependences of muscle fibers or muscles separately from the elastic properties of ligaments, tendons, aponeuroses and other connecting tissue. When speaking of Hill-type muscle models, nowadays a combination of a contractile element (CE) describing merely contractile properties, and different sorts of elastic elements (linear, quadratic, ...), added in parallel (PEE) and/or in series (SEE), is meant. In addition, also parallel and/or serial damping elements (PDE, SDE) and other mechanically describable elements can be added. The single elements can be combined in various configurations (Haeufle et al., 2014; Hof, 1998; Mörl, Siebert, Schmitt, Blickhan & Günther, 2012; Piovesan, Pierobon & Mussa Ivaldi, 2013; Rockenfeller & Günther, 2017; Siebert, Rode, Herzog, Till & Blickhan, 2008; Thelen, 2003; Winters, 1990; Yu & Wilson, 2014). For linear elastic elements, some of these configurations are mathematically identical but one has to be very careful about the order when combining nonlinear elements (see, e.g., Hof, 2003; Siebert et al., 2008). Figure 2.6 shows two of the most commonly used configurations for contractile and elastic elements.

Modelling the whole-muscle force, one has to be aware that the direction of force may be different from the direction of the force acting in the fibers. Depending on the kind of muscle (unipennate, multipennate), there are one or more pennation angles to be considered when calculating the resultant force (e.g., Maganaris, Baltzopoulos & Sargeant, 1998; Woittiez et al., 1984).

For geometrical reasons, the properties of tendons are commonly described by an SEE, whereas the aponeurosis acts more as a PEE. However, the SEE stiffness supports the view that the characteristics of the SEE are more similar to the characteristics of the aponeurosis than to those of the tendon (Epstein, Wong & Herzog, 2006; Penasso & Thaller, 2017; Tilp, Steib & Herzog, 2012).

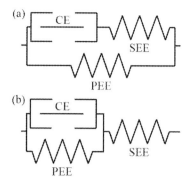

Figure 2.6 Two commonly used combinations of a contractile element together with parallel and serial elastic elements: (a) the serial elastic element (SEE) is attached to the contractile element (CE), whereas the parallel elastic element (PEE) acts over the whole length, (b) the SEE is attached to the combination of CE and PEE. Due to nonlinearities, versions (a) and (b) do not reflect the same situation.

Stretch-shortening cycle

During most human movements, muscles act in concentric, isometric and eccentric ways alternately. In the so-called stretch-shortening cycle (SSC), a concentric movement immediately follows an eccentric stretch that leads to a performance enhancement. A typical example for such a movement is the countermovement jump. The observed increase in muscle performance of an SSC is ascribed to, e.g., a higher activation level, elastic energy storing and release and residual force enhancement (Van Ingen Schenau, Bobbert & De Haan,1997a, 1997b; Komi, 2000; Seiberl, Power, Herzog & Hahn, 2015). Thus, at least some of these effects have to be included in models for movements with SSCs.

Results

We have discussed force laws for the muscle (i.e., Hill's equation including length and velocity dependence, combined with serial and parallel components for elasticity and viscosity). However, Hill's equation is only valid if the muscle is fully activated. For getting the effective muscle force, the equation of motion has also to incorporate muscle activation dynamics, which also has to be formulated using movement-independent parameters. Furthermore, the geometric situation has to be taken into account to describe how the muscle force is transferred to the body parts under consideration. Finally, external forces like gravitation, reaction forces or inertia forces must be included. Depending on the modelling goal, further details may be added to the model. Models for simulation may be very complex. In contrast to these models for simulation, models for finding muscular parameter values using inverse dynamics must not be underdetermined. Nevertheless, they can also be used for simulation of simple movements.

In vivo determination of Hill's parameters

In order to determine individual neuromuscular parameters of the knee extensors non-invasively, we present a simple model of a knee extension movement on an inclined leg press. The model originates from Sust (1987), and improvements were added in (Penasso, 2014; Penasso & Thaller, 2017; Siebert et al., 2007). Different versions of the model were used to identify Hill's parameters, elastic properties, and activation rates (Penasso & Thaller, 2017; Siebert et al., 2007; Thaller & Wagner, 2004; Wagner, Siebert, Ellerby, Marsh & Blickhan, 2005; Wagner, Thaller, Dahse & Sust, 2006). It was used to investigate the relation between muscle properties and genetics, the influence of training interventions, fiber distribution, fatigue, effect of caffeine consumption, muscle stress and many further questions. Data on individual parameters can be found in the papers mentioned.

Consider an inclined leg press with an inclination angle α. The subjects push a load m by extending their legs. The position of the load is denoted by X, the distance between the COM of the load and the hip joint. The velocity of the load is denoted by $V = \dot{X}$. The forces that act on the load are the force component of the weight against movement direction and the effective muscle force transferred to the load (see Figure 2.7). Inserting this into Newton's equation we get the following differential equation for the movement of the COM:

$$m \cdot \ddot{X} = -m \cdot g \cdot \sin \alpha + A(t) \cdot G(X) \cdot \left(\frac{c}{G(X) \cdot \dot{X} + b} - a \right) \qquad (30)$$

where g denotes the gravitational acceleration, a, b and c are Hill's constants for a model muscle that acts as a knee extensor. $A(t)$ describes the activation of a

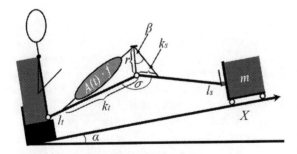

Figure 2.7 Knee extension movement on a leg press with inclination angle α. X denotes the position of the mass pushed along an inclined axis. The knee angle σ is calculated via the length of the thigh l_t, the length of the shank l_s, and X. The figure shows further parameters used in equations (32)–(34) for calculating the transfer of the muscle force to the force acting on the mass (see also text).

maximum voluntary contraction, using the movement-independent parameter S and the shift parameter τ,

$$A(t) = 1 - \exp(-S(t - \tau)). \tag{31}$$

$G(X)$ is the transfer function that relates the effective muscle force $A(t) \cdot f$, $f = c/(v+b) - a$, to the force acting on the COM

$$G(X) = \frac{r \sin \beta}{l_t l_s \sin \sigma} \cdot X \tag{32}$$

with

$$\sigma = 2\beta + \arcsin\left(\frac{r}{k_t} \sin \beta\right) + \arcsin\left(\frac{r}{k_s} \sin \beta\right) \text{ and} \tag{33}$$

$$X = \sqrt{l_t^2 + l_s^2 - 2 l_t l_s \cos \sigma}. \tag{34}$$

The length of the thigh l_t, the length of the shank l_s, the positions of insertion of the model muscle k_t, k_s and the knee radius r are measured directly. The angles σ (knee extension angle) and β (angle between force directions within the knee model) are calculated according to the above equations. A detailed description of the knee model can be found in Sust, Schmalz, Beyer, Rost, Hansen and Weiss (1997a).

The initial velocity $V(0)$ is set zero, and thus the load is at rest at $t = 0$. τ is the shift of the activation function according to the equilibrium condition

$$A(0) = \frac{m \cdot g \cdot \sin \alpha}{G(X(0)) \cdot f_{iso}}. \tag{35}$$

Therefore,

$$\tau = \frac{1}{S} \ln(1 - \frac{m \cdot g \cdot \sin \alpha \cdot b}{G(X(0)) \cdot (c - ab)}). \tag{36}$$

In the model, m, g, α, the initial position $X(0)$, and $V(0)$ are conditions of the movement; all other parameters are properties of the subject. As the mass m of the load and the value of g are given, only a, b, c and S have to be identified.

These basic modelling thoughts were used for identifying muscle properties in vivo. A very detailed description of the identification process and an improved model including a more elaborate activation function and elastic components can be found in Penasso and Thaller (2017). One has to be aware that different models lead to slightly different parameter values. For example, if the model does

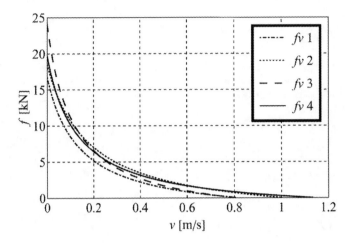

Figure 2.8 Individual force-velocity relations fv 1–fv 4 using the identified parameter values (2826, 0.139, 2714), (3288, 0.180, 3922), (3398, 0.114, 3128) and (2257, 0.131, 2877), respectively, in SI-units for a, b and c in Hill's Equation (9). Some curves intersect, showing that individual differences depend on the velocity: larger isometric force does not imply larger force at higher velocities.

Note: data from Penasso and Thaller (2017)

not include elastic properties (Siebert et al., 2007, 2008), the effect of elasticity is included in the identified parameters of the CE. However, one can minimize these differences by choosing appropriate movements for the identification process (slow vs. fast movements, small vs. high loads, large vs. small inclination angle …).

Figure 2.8 shows individual differences in four force-velocity relations. The identified four parameter sets for a [N], b [m/s] and c [W] are (2826, 0.139, 2714), (3288, 0.180, 3922), (3398, 0.114, 3128) and (2257, 0.131, 2877) for the four curves fv 1, fv 2, fv 3 and fv 4, respectively (Penasso & Thaller, 2017). The graphs have different curvatures; the curves intersect, some even twice. That shows that muscles that are strong in slow movements need not be strong in faster movements (see also Exercise 1).

Sensitivity analysis

Sensitivity analysis leads to interesting applications in exercise science: In many sports, the athlete's performance is assessed by a real number, e.g., the height in a jump or the time in a running competition. If we have a proper structural model of the movement, the performance can be calculated using the parameters of the model equations, i.e., the conditions of the movement and the state of the person. A performance-determining factor is a parameter that has a large influence on the performance, independent of the other parameter values. However, in some parameter value constellations, a parameter may also be performance limiting, i.e., whatever value the other parameters might have, the performance does not increase unless this parameter is changed. Thus, the value of a performance-limiting parameter has to be changed in order to get any increase in the performance. If we define a function mapping the state of the person to the performance, the largest slope of the graph, the gradient, indicates the path of quickest increase in performance. On the other hand, when performing this movement as an exercise, the gradient may indicate the combination of parameter value changes. The gradient depends crucially on the individual parameters. A more detailed mathematical formulation is presented in Thaller, Tilp and Sust (2010).

Dimensionless parameters: Curvature of the force-velocity relation and muscle efficiency

Dimensionless parameters are sometimes very useful because they do not depend on the specific unit system. Furthermore, they retain the same value after normalization to muscle length, muscle cross section or maximum force.

In the equations dealing with the length dependence of the muscle force (equations (1)–(3)), the force at length l is given in relation to the force at l_0, thus acting as a dimensionless factor independent of the units used.

According to the Buckingham π theorem (Buckingham, 1914), there is at least one independent dimensionless parameter for Hill's Equation (9). For example, the number ab/c has dimension 1. It correlates with the curvature of the force-velocity function. The ratio between a and f_{iso}, a/f_{iso}, is another dimensionless parameter, but it is not independent of ab/c. In fact, there is a strict relation to ab/c, and a short calculation shows that a/f_{iso} can be expressed by

$$\frac{a}{f_{iso}} = \frac{\frac{ab}{c}}{1 - \frac{ab}{c}}. \tag{37}$$

Hill (1938) argued that the ratio between the parameter a, which denotes some friction in the muscle, and the isometric force f_{iso}, should be constant. For the frog sartorius muscle, he found $a/f_{iso} = 0.25$ (Hill, 1938, p. 160) and $a/f_{iso} = 0.22$ (Hill, 1938, p. 177). Epstein and Herzog (1998) propose $a/f_{iso} = 0.25$. This implies $ab/c = 0.2$. Further average values are used in literature, e.g., $a/f_{iso} = 0.41$ (Van Soest & Bobbert, 1993).

In a similar way, we get a strict relation to the curvature for $\frac{f_{opt}}{f_{iso}}$ and $\frac{v_{opt}}{v_{max}}$,

$$\frac{f_{opt}}{f_{iso}} = \frac{v_{opt}}{v_{max}} = \frac{-\frac{ab}{c} + \sqrt{\frac{ab}{c}}}{1 - \frac{ab}{c}}. \tag{38}$$

In contrast to $f_{opt}/f_{iso} = v_{opt}/v_{max}$, Zatsiorsky and Kraemer (2006) used $f_{opt}/f_{iso} = 1/2$, and $v_{opt}/v_{max} = 1/3$.

If we define the efficiency η by the relation between the mechanical power p and the chemical power c, $\eta = p/c$, (Wagner et al., 2006), we get

$$\eta_{max} = \frac{p_{max}}{c} = (1 - \sqrt{\frac{ab}{c}})^2. \tag{39}$$

For other ways of defining the efficiency, see, e.g., Van Ingen Schenau, Bobbert and De Haan (1997b).

From Equation (12) we immediately see that $\eta_{max} < 1$. For $v = 0$ and $v = v_{max}$, the mechanical power is zero and therefore the efficiency is also zero. This implies that all chemical energy per time is converted either into heat, into mechanical work on a lower hierarchical level or into potential energy stored in elastic structures. Fenn (1923) already observed that a muscle produces more heat during an isometric movement than during a concentric movement over the same time. As already mentioned, the rate of heat production $a \cdot v$ stated by Hill (1938) does not mean that there is no heat produced at zero velocity. For $a/f_{iso} = 0.25$ we get $\eta_{max} = 0.3$.

Composing equal contracting elements in parallel and serial arrangements yields structures with the same curvature and efficiency. From Equation (26) we get

$$\frac{a_p b_p}{c_p} = \frac{(2a) \cdot b}{2c} = \frac{ab}{c}. \tag{40}$$

However, there are inter-individual differences in the muscle efficiency and the curvature of Hill's equation, and therefore a/f_{iso} is not a constant for all muscles. Zatsiorsky and Kraemer (2006) distinguished between athletes in power sports with $a/f_{iso} > 0.3$ on one side and endurance athletes and beginners with $a/f_{iso} < 0.3$ on the other side. That implies that power athletes are less efficient than endurance athletes, who have more slow-twitch fibers. Comparing marathon runners to high jump athletes shows that jumpers perform their action over a very short period and thus need not be energy-efficient. Furthermore, fiber types differ in p_{max} and v_{max} (Bottinelli, Canepari, Pellegrino & Reggiani, 1996; Bottinelli, Schiaffino & Reggiani, 1991; Brooks & Faulkner, 1991; Linari, Bottinelli, Pellegrino, Reconditi, Reggiani & Lombardi, 2004; MacIntosh et al., 1993). Fast-twitch fibers have higher maximum velocity than slow-twitch fibers with the same isometric force. Sust et al. (1997b) showed that there is also a correlation ($r = 0.92$, $p < 0.05$) between the percentage of fast fibers and the value of b_n, where b_n denotes Hill's constant b scaled to the muscle length. Therefore, there is not only a difference in efficiency but also in the velocity of the chemical reactions in the muscle, reflected by b. Measurements on the knee extensor muscles of 62 male and female sports students show that there is a strong correlation between a/f_{iso} and b_n, $r = 0.91$, $p < 0.05$ (Thaller & Wagner, 2004).

Examples

Phenomena (partly) caused by the velocity dependence of muscle force

There are many phenomena observed in human movements that can be explained—at least partly—by the velocity dependence of the muscle force. Simulations show that individual differences stem (partly) from differences in the individual force-velocity relations of the muscles.

Arm movement in a vertical jump and in the ancient Olympic long jump

Why do we use our arms when jumping? In a vertical jump without countermovement, the upward movement of the arms starts approximately at the same time as extending the legs. Accelerating the arms upwards yields a vertical force downwards acting against the extension of the legs. Therefore, the extension is slower in the early phase of the jump. The muscles can produce more force at slower velocity and thus the integral of the force over the time, the change in momentum, is larger. Normally, the arms decelerate during the flight phase, which leads to a further increase of the jump height, i.e., by transfer of momentum.

The same mechanism was utilized by the athletes in the ancient Olympic long jump: They held additional weight in their hands to increase their performance (Minetti & Ardigó, 2002). Differences in the optimum weight can be explained by individual differences in the force-velocity relation (Thaller, Sust & Tilp, 2003).

Bilateral deficit

Performing an isometric leg extension in a bench press with both legs yields less force than the sum of both single leg extensions. This bilateral deficit has been often ascribed to vaguely defined neural activation processes (for review, see Škarabot, Cronin, Strojnik and Avela, 2016). For concentric movements, a major part of this phenomenon can be ascribed to the force-velocity relation. Simulating a leg extension using the same activation properties for one and for two legs shows a non-proportional difference between the unilateral and the bilateral extension because the unilateral extension is slower. Thus, the unilateral movement produces higher forces over a longer period compared to the force in the bilateral extension (Bobbert, De Graaf, Jonk & Casius, 2006).

Muscle stress

Externally measured quantities are often used to individualize training programs. They are a mixture of movement conditions and individual properties, and depend on the activation dynamics as well as on the geometrical situation and the dynamics of the muscle-tendon complex. For example, the commonly used one repetition maximum (1RM) depends on the joint angle. Furthermore, as the contraction velocity at a 1RM exercise is very low, only a small part of the force-velocity relation is assessed. Nevertheless, the 1RM is often used to determine the load for training movements with faster velocities. Thus, depending on the individual curvature of the force-velocity relation, muscles of two subjects with equal 1RMs but different a/f_{iso} are differently stressed during the resulting dynamic training movements (Penasso, Binder & Thaller, 2013; Thaller & Penasso, 2013).

Exercises

1) For two fictitious muscles, let a_1 = 5000 N, b_1 = 0.1 m/s, c_1 = 4500 W and a_2 = 3000 N, b_2 = 0.15 m/s, c_2 = 4000 W be the values in Hill's Equation (9). Using equations (10), (11) and (16), calculate f_{iso}, v_{max} and p_{max} for both muscles. Which muscle is stronger at v = 0 m/s, and which at v = 0.4 m/s? Which muscle is more efficient (use Equation (39))? At which velocities do the muscles exert maximal mechanical power (use Equation (14))?

2) Inspired by Hill's experiments in 1938, write a simple model for a fully activated muscle that works against a constant load (quick release protocol as in the paper) using a massless muscle without any elastic components (Figure 2.9). The CE should contain a) only a force-velocity relation (Equation (9)) and b) also a force-length relation (Equation (1)). What load-velocity relation do you get? Keep in mind that we have Newton's equation of motion, i.e., we have to consider all the forces that act on the load. The forces are the muscle force acting in one direction (Equation (9)) and the load acting in the opposite direction. In question b), the muscle force also has to be multiplied by Equation (3). When the muscle force is equal to the (negative) load, we have equilibrium and the velocity stays constant.

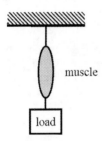

Figure 2.9 Sketch of the situation in example 2. The fixed load is released as soon as the muscle is fully activated. Two forces act on the load: the force exerted by the contracting muscle and the load (mass × gravitational acceleration).

3) Consider the equation of motion (30) for an extension movement on an inclined leg press. The load is first held at the initial position X_0. Then the movement starts by extending the legs with MVC.

a) How does the initial position influence the maximal power $\max_t p(t)$? Keep in mind that the initial activation depends on the initial position and that $\max_t p(t)$ depends on the activation of the muscle at that time.

b) How does the mass influence the initial activation? Consider that not only the initial position influences the initial activation, but also the mass that has to be held at the beginning of the movement.

To help answering the questions, a Matlab® function (KneeExt.m) for simulation can be found on the MathWorks® home page (Penasso & Thaller, 2018).

References

Anderson, F. & Pandy, M. (2001). Dynamic optimization of human walking. *Journal of Biomechanical Engineering*, 123(5), 381–390.

Bayer, A., Schmitt, S., Günther, M. & Haeufle, D. (2017). The influence of biophysical muscle properties on simulating fast human arm movements. *Computer Methods in Biomechanics and Biomedical Engineering*, 20(8), 803–821.

Bobbert, M. (2012). Why is the force-velocity relationship in leg press tasks quasi-linear rather than hyperbolic? *Journal of Applied Physiology*, 112(12), 1975–1983.

Bobbert, M., De Graaf, W., Jonk, J. & Casius, L. (2006). Explanation of the bilateral deficit in human vertical squat jumping. *Journal of Applied Physiology*, 100(2), 493–499.

Bottinelli, R., Canepari, M., Pellegrino, M. & Reggiani, C. (1996). Force-velocity properties of human skeletal muscle fibres: myosin heavy chain isoform and temperature dependence. *The Journal of Physiology*, 495(2), 573–586.

Bottinelli, R., Schiaffino, S. & Reggiani, C. (1991). Force-velocity relations and myosin heavy chain isoform compositions of skinned fibres from rat skeletal muscle. *The Journal of Physiology*, 437: 655–672.

Brooks, S. & Faulkner, J. (1991). Forces and powers of slow and fast skeletal muscles in mice during repeated contractions. *The Journal of Physiology*, 436: 701–710.

Buckingham, E. (1914). On physically similar systems: Illustrations of the use of dimensional equations. *Physical Review*, 4(4), 345–376.

Burkholder, T. & Lieber, R. (2001). Sarcomere length operating range of vertebrate muscles during movement. *The Journal of Experimental Biology*, 204(9), 1529–1536.

Crowninshield, R. & Brand, R. (1981). A physiologically based criterion of muscle force prediction in locomotion. *Journal of Biomechanics*, 14(11), 793–801.

Epstein, M. & Herzog, W. (1998). *Theoretical Models of Skeletal Muscle: Biological and Mathematical Considerations*, New York: Wiley.

Epstein, M., Wong, M. & Herzog, W. (2006). Should tendon and aponeurosis be considered in series? *Journal of Biomechanics*, 39(11), 2020–2025.

Fenn, W. (1923). A quantitative comparison between the energy liberated and the work performed by the isolated sartorius muscle of the frog. *The Journal of Physiology*, 58 (2–3), 175–203.

Fenn, W. & Marsh, B. (1935). Muscular force at different speeds of shortening. *The Journal of Physiology*, 85(3), 277–297.

Gasser, H. & Hill, A. (1924). The dynamics of muscular contraction. In *Proceedings of the Royal Society B: Biological Sciences*, 96(678), 398–437.

Gordon, A., Huxley, A. & Julian, F. (1966). The variation in isometric tension with sarcomere length in vertebrate muscle fibres. *The Journal of Physiology*, 184(1), 170–192.

Haeufle, D., Günther, M., Bayer, A. & Schmitt, S. (2014). Hill-type muscle model with serial damping and eccentric force-velocity relation. *Journal of Biomechanics*, 47(6), 1531–1536.

Hatze, H. (1976). The complete optimization of a human motion. *Mathematical Biosciences*, 28(1–2), 99–135.

Hatze, H. (1977). A myocybernetic control model of skeletal muscle. *Biological Cybernetics*, 25(2), 103–119.

Heidlauf, T., Klotz, T., Rode, C., Siebert, T. & Röhrle, O. (2017). A continuum-mechanical skeletal muscle model including actin-titin interaction predicts stable contractions on the descending limb of the force-length relation. *PLoS Computational Biology*, 13(10), e1005773.

Herzog, W. (Ed.). (2000) *Skeletal muscle mechanics: From mechanisms to function*. New York: Wiley.

Herzog, W. (2014). Mechanisms of enhanced force production in lengthening (eccentric) muscle contractions. *Journal of Applied Physiology*, 116(11), 1407–1417.

Herzog, W. & Leonard, T. (2002). Force enhancement following stretching of skeletal muscle: A new mechanism. *The Journal of Experimental Biology*, 205(9), 1275–1283.

Hill, A. (1938). The heat of shortening and the dynamic constants of muscle. In *Proceedings of the Royal Society B: Biological Sciences*, 126(843), 136–195.

Hill, A. (1970). *First and last experiments in muscle mechanics*. Cambridge: Cambridge University Press.

Hof, A. (1998). In vivo measurement of the series elasticity release curve of human triceps surae muscle. *Journal of Biomechanics*, 31, 793–800.

Hof, A. (2003). Muscle mechanics and neuromuscular control. *Journal of Biomechanics*, 36(7), 1031–1038.

Hoy, M., Zajac, F. & Gordon, M. (1990). A musculoskeletal model of the human lower extremity: The effect of muscle, tendon, and moment arm on the moment-angle relationship of musculotendon actuators at the hip, knee, and ankle. *Journal of Biomechanics*, 23(2), 157–169.

Huxley, A. (1957). Muscle structure and theories of contraction. *Progress in biophysics and biophysical chemistry*, 7, 255–318.

Huxley, A. & Niedergerke, R. (1954). Structural changes in muscle during contraction: Interference microscopy of living muscle fibres. *Nature*, 173(4412), 971–973.

Huxley, A. & Simmons, R. (1971). Proposed mechanism of force generation in striated muscle. *Nature*, 233(5321), 533–538.

Huxley, H. & Hanson, J. (1954). Changes in the cross-striations of muscle during contraction and stretch and their structural interpretation. *Nature*, 173(4412), 973–976.

Katz, B. (1939). The relation between force and speed in muscular contraction. *The Journal of Physiology*, 96(1), 45–64.

Komi, P. (2000). Stretch-shortening cycle: A powerful model to study normal and fatigued muscle. *Journal of Biomechanics*, 33(10), 1197–1206.

Leonard, T., DuVall, M. & Herzog, W. (2010). Force enhancement following stretch in a single sarcomere. *American Journal of Physiology. Cell Physiology*, 299(6), C1398–1401.

Linari, M., Bottinelli, R., Pellegrino, M., Reconditi, M., Reggiani, C. & Lombardi, V. (2004). The mechanism of the force response to stretch in human skinned muscle fibres with different myosin isoforms. *The Journal of Physiology*, *554*(2), 335–352.

MacIntosh, B., Herzog, W., Suter, E., Wiley, J. & Sokolosky, J. (1993). Human skeletal muscle fibre types and force: Velocity properties. *European Journal of Applied Physiology and Occupational Physiology*, *67*(6), 499–506.

MacIntosh, B., Gardiner, P. & McComas, A. (2016) *Skeletal Muscle* (2nd edn.). Champaign, IL: Human Kinetics.

Maganaris, C., Baltzopoulos, V. & Sargeant, A. (1998). In vivo measurements of the triceps surae complex architecture in man: Implications for muscle function. *The Journal of Physiology*, *512* (2), 603–614.

Minetti, A. & Ardigó, L. (2002). Halteres used in ancient Olympic long jump. *Nature*, *420*(6912), 141–142.

Mitsui, T. & Ohshima, H. (1988). A self-induced translation model of myosin head motion in contracting muscle. I. Force-velocity relation and energy liberation. *Journal of Muscle Research and Cell Motility*, *9*(3), 248–260.

Mörl, F., Siebert, T., Schmitt, S., Blickhan, R. & Günther, M. (2012). Electro-mechanical delay in Hill-type muscle models. *Journal of Mechanics in Medicine and Biology*, *12*(5), 1250085.

Nigg, B. & Herzog, W. (Eds.). (1999). *Biomechanics of the musculo-skeletal system* (2nd edn.). New York: Wiley.

Pandy, M., Zajac, F., Sim, E. & Levine, W. (1990). An optimal control model for maximum-height human jumping. *Journal of Biomechanics*, *23*(12), 1185–1198.

Penasso, H. Activating muscles from pre-activation to MVC. Paper presented at The 19th Annual Congress of the European College of Sport Science, Amsterdam, July 2014.

Penasso, H., Binder, I. & Thaller, S. Individually shaped force-velocity relations effect muscle stress during isokinetic leg press movements with MVC. Paper presented at The 18th Annual Congress of the European College of Sport Science, Barcelona, June 2013.

Penasso, H. & Thaller, S. (2017). Determination of individual knee-extensor properties from leg extensions and parameter identification. *Mathematical and Computer Modeling of Dynamical Systems*, *23*(4), 416–438.

Penasso, H. & Thaller, S. (2018). *KneeExt.m*', *MathWorks*®, http://de.mathworks.com/matlabcentral/fileexchange/65975-kneeext, online February 7, 2018.

Piovesan, D., Pierobon, A. & Mussa Ivaldi, F. (2013). Critical damping conditions for third order muscle models: Implications for force control. *Journal of Biomechanical Engineering*, *135*(10), 101010.

Podolsky, R. (1960). The structure and function of muscle. In G. Bourne (Ed.), *The structure and function of muscle* (2nd edn.). London: Academic Press.

Podolsky, R. (1961). The mechanism of muscular contraction. *The American Journal of Medicine*, *30*(5), 708–719.

Podolsky, R., Nolan, A. & Zaveler, S. (1969). Cross-bridge properties derived from muscle isotonic velocity transients. *Proceedings of the National Academy of Sciences of the United States of America*, *64*(2), 504–511.

Rajagopal, A., Dembia, C., DeMers, M., Delp, D., Hicks, J. & Delp, S. (2016). Full-body musculoskeletal model for muscle-driven simulation of human gait. *IEEE Transactions on Biomedical Engineering*, *63*(10), 2068–2079.

Rassier, D., Herzog, W. & Pollack, G. (2003). Dynamics of individual sarcomeres during and after stretch in activated single myofibrils. *Proceedings of the Royal Society of London B: Biological Sciences*, *270*(1525), 1735–1740.

Rockenfeller, R. & Günther, M. (2017). How to model a muscle's active force-length relation: A comparative study. *Computer Methods in Applied Mechanics and Engineering*, *313*, 321–336.

Rockenfeller, R., Günther, M., Schmitt, S. & Götz, T. (2015). Comparative sensitivity analysis of muscle activation dynamics. *Computational and Mathematical Methods in Medicine*, *2015*(585409), 1–16.

Rode C, Siebert T, Tomalka, A & Blickhan R. (2016). Myosin filament sliding through the Z-disc relates striated muscle fibre structure to function. *Proceedings of the Royal Society of London B: Biological Sciences*, *283*(1826), 20153030.

Samozino, P., Morin, J., Hintzy, F. & Belli, A. (2010). Jumping ability: A theoretical integrative approach. *Journal of Theoretical Biology*, *264*(1), 11–18.

Schappacher-Tilp, G., Leonard, T., Desch, G. & Herzog, W. (2015). A novel three-filament model of force generation in eccentric contraction of skeletal muscles. *PLoS One*, *10*(3), e0117634.

Scovil, C. & Ronsky, J. (2006). Sensitivity of a Hill-based muscle model to perturbations in model parameters. *Journal of Biomechanics*, *39*(11), 2055–2063.

Seiberl, W., Power, G., Herzog, W. & Hahn, D. (2015). The stretch-shortening cycle (SSC) revisited: Residual force enhancement contributes to increased performance during fast SSCs of human m. adductor pollicis. *Physiological Reports*, *3*(5), 1–12.

Siebert, T., Rode, C., Herzog, W., Till, O. & Blickhan, R. (2008). Nonlinearities make a difference: Comparison of two common Hill-type models with real muscle. *Biological Cybernetics*, *98*(2), 133–143.

Siebert, T., Sust, M., Thaller, S., Tilp, M. & Wagner, H. (2007). An improved method to determine neuromuscular properties using force laws: From single muscle to applications in human movements. *Human Movement Science*, *26*(2), 320–341.

Simmons, R. (Ed.). (1992). *Muscular contraction*. Cambridge: Cambridge University Press.

Škarabot, J., Cronin, N., Strojnik, V. & Avela, J. (2016). Bilateral deficit in maximal force production. *European Journal of Applied Physiology*, *116*(11–12), 2057–2084.

Stockhecker, H. (1995). Theoretische und experimentelle Untersuchung zu Kraftgesetzen für die Beschreibung menschlicher Muskelkontraktionen [Theoretical and experimental investigations on force laws to describe human muscle contractions]. Unpublished Ph.D. thesis, Johann Wolfgang Goethe-Universität, Frankfurt.

Sust, M. (1978). Biomechanische Aspekte der Definition von Maximal-und Schnellkraft [Biomechanical aspects of the definition of maximum force and rate of force development]. *Theorie und Praxis der Körperkultur*, *27*(10), 763–768.

Sust, M. (1987). Beitrag zum Aufbau einer axiomatischen Theorie der Biomechanik und Beispiele ihrer Anwendung [Contribution to the development of an axiomatic theory of biomechanics and examples of its application]. Unpublished habilitation thesis, Friedrich-Schiller Universität, Jena, Germany.

Sust, M. (2009). Hillsche Gleichung und die Sportwissenschaft: 70 Jahre Hillsche Gleichung [Hill's equation and sport science: 70 years of Hill's equation]. *Spectrum*, *21*(1), 38–67.

Sust, M., Schmalz, T., Beyer, L., Rost, R., Hansen, E. & Weiss, T. (1997a). Assessment of isometric contractions performed with maximal subjective effort: Corresponding results for EEG changes and force measurements. *The International Journal of Neuroscience*, *92*(1–2), 103–118.

Sust, M., Schmalz, T. & Linnenbecker, S. (1997b). Relationship between distribution of muscle fibres and invariables of motion. *Human Movement Science*, *16*(4), 533–546.

Thaller, S. & Penasso, H. (2013). The influence of Hill's force-velocity relation on muscle stress during exercises with constant load and velocity. Paper presented at The 18th Annual Congress of the European College of Sport Science, Barcelona, June 2013.

Thaller, S., Sust, M. & Tilp, M. (2003). Individual effects of additional weights used in ancient Greek Olympic pentathlon. Paper presented at The 8th Annual Congress of the European College of Sport Science, Salzburg, July 2003.

Thaller, S., Tilp, M. & Sust, M. (2010). The effect of individual neuromuscular properties on performance in sports. *Mathematical and Computer modeling of Dynamical Systems*, 16(5), 417–429.

Thaller, S. & Wagner, H. (2004). The relation between Hill's equation and individual muscle properties. *Journal of Theoretical Biology*, 231(3), 319–332.

Thelen, D. (2003). Adjustment of muscle mechanics model parameters to simulate dynamic contractions in older adults. *Journal of Biomechanical Engineering*, 125(1), 70–77.

Tilp, M., Steib, S. & Herzog, W. (2012). Length changes of human tibialis anterior central aponeurosis during passive movements and isometric, concentric, and eccentric contractions. *European Journal of Applied Physiology*, 112(4), 1485–1494.

Van Ingen Schenau, G., Bobbert, M. & De Haan, A. (1997a). Does elastic energy enhance work and efficiency in the stretch-shortening cycle? *Journal of Applied Biomechanics*, 13(4), 389–415.

Van Ingen Schenau, G., Bobbert, M. & De Haan, A. (1997b). Mechanics and energetics of the stretch-shortening cycle: A stimulating discussion. *Journal of Applied Biomechanics*, 13(4), 484–496.

Van Soest, A. & Bobbert, M. (1993). The contribution of muscle properties in the control of explosive movements. *Biological Cybernetics*, 69(3), 195–204.

Van Soest, A., Schwab, A., Bobbert, M. & Van Ingen Schenau, G. (1993). The influence of the biarticularity of the gastrocnemius muscle on vertical-jumping achievement. *Journal of Biomechanics*, 26(1), 1–8.

Wagner, H., Siebert, T., Ellerby, D., Marsh, R. & Blickhan, R. (2005). ISOFIT: A model-based method to measure muscle-tendon properties simultaneously. *Biomechanics and Modeling in Mechanobiology*, 4(1), 10–19.

Wagner, H., Thaller, S., Dahse, R. & Sust, M. (2006). Biomechanical muscle properties and angiotensin-converting enzyme gene polymorphism: A model-based study. *European Journal of Applied Physiology*, 98(5), 507–515.

Winby, C., Lloyd, D. & Kirk, T. (2008). Evaluation of different analytical methods for subject-specific scaling of musculotendon parameters. *Journal of Biomechanics*, 41(8), 1682–1628.

Winters, J. (1990). Hill-based muscle models: A systems engineering perspective. In J. Winters & S. Woo (Eds.), *Multiple Muscle Systems*. New York: Springer.

Winters, J. (1995). An improved muscle-reflex actuator for use in large-scale neuromusculoskeletal models. *Annals of Biomedical Engineering*, 23(4), 359–374.

Woittiez, R., Huijing, P., Boom, H. & Rozendal, R. (1984). A three-dimensional muscle model: A quantified relation between form and function of skeletal muscles. *Journal of Morphology*, 182(1), 95–113.

Yamaguchi, G. & Zajac, F. (1989). A planar model of the knee joint to characterize the knee extensor mechanism. *Journal of Biomechanics*, 22(1), 1–10.

Yu, T. & Wilson, A. (2014). A passive movement method for parameter estimation of a musculo-skeletal arm model incorporating a modified Hill muscle model. *Computer Methods and Programs in Biomedicine.* 114(3), e46–59.

Zajac, F. (1989). Muscle and tendon: Properties, models, scaling, and application to biomechanics and motor control. *Critical Reviews in Biomedical Engineering*, 17(4), 359–411.

Zatsiorsky, V. & Kraemer, W. (2006) *Science and practice of strength training* (2nd edn.). Champaign, IL: Human Kinetics.

3 Rowing

Michał Wychowański and Arnold Baca

Introduction

The art of propelling a boat on water by the use of oars originated a very long time ago. Probably, a broken bough was the first tool used to propel a tree trunk floating in water. After many years, the bough has taken the form of a wooden paddle. The first crafts to be propelled with paddles were canoes. Kayaks were created thousands of years ago by the Eskimos in the northern regions of the Arctic. The main use of kayaks was to hunt and fish. The first kayaks, like today's tourist and sport kayaks, were driven by a double-blade paddle held in two hands. A canoe was a North American indigenous people paddling boat. Europeans used those boats to explore the new continent. Canoes were driven with a single-blade paddle kept in two hands. The earliest boats, for which paddling played an important role, were the ones driven by the slaves of Egyptian, Roman or Arab rulers. In ancient times, paddle driving was of paramount importance during sea battles.

Currently, paddles are used in sport and competition. Canoeing is an athletic or recreational activity that involves paddling a canoe with a single-blade paddle. Kayaking is a sport of driving a small, narrow watercraft with a double-blade paddle.

In rowing, two variants are distinguished: proper rowing and sculling. Rowing is propelling the boat with one oar held in both hands; in sculling, the oars propel the boat by moving through the water on both sides of the craft. In a single scull, one rower propels the boat with two oars.

The common feature of kayaking, canoeing, rowing and sculling is how to develop power by propelling the boat through the water using a paddle or an oar. The sports rowing and sculling oar is anchored to the boat by the oarlock. The oarlock allows the oar's blade to be moved under the surface of the water in a vertical position during the stroke as well as to be turned to a horizontal position during the recovery phase, until the next water catch. The generation of driving force with the oar is a complicated process, difficult for a mathematical description but necessary to model the rowing motion.

The methodological principles of mathematical modelling and measurements used in sport have significantly influenced the intensive development of all branches of biomechanics during the period of its dynamic progression in the 1970s. Perhaps, one of the best definitions of biomechanics was provided by Herbert Hatze in 1974: "Biomechanics is the study of the structure and function of biological systems by means of the methods of mechanics." The word "biomechanics" developed to describe the application of engineering mechanics to the biological and medical systems (Hatze, 1974).

In Poland, Adam Morecki, Juliusz Ekiel and Kazimierz Fidelus published a very important book entitled *Bionics of Motion* that has formed the foundations of modern biomechanics (Morecki, Ekiel & Fidelus, 1971). The main achievement of the authors was the creation of a kinematic model of human structures including 240 degrees of freedom on the basis of the theory of machines and mechanisms. Herbert Hatze published the book *Myocybernetic Control Models of Skeletal Muscle* that opened up the era of cybernetic modelling in biomechanics and muscle control (Hatze, 1981).

In 1976, the Institute of Sport was established in Warsaw. The main statutory goal of the institute's operation was to support Polish high-performance athletes. Special methodological teams composed of scientists and specialists in various fields were involved in preparing athletes for competition. At that time, great attention was paid to support canoeing and rowing. The results of that activity were numerous scientific studies and even patents (Wychowański, Nosarzewski & Karpiłowski, 1987a; Wychowański, Nosarzewski & Witkowski, 1987b; Wychowański et al., 1989).

An established way of obtaining information about a dynamical process is to measure directly the parameters and functions describing the dynamics of this process. Actually, the measurements during training on water or on special ergometers in the laboratory are commonly performed (Torres-Moreno, Tanaka & Penney, 2000; Hofmann et al., 2007; Dworak, 2010; Findlay & Turnock, 2010; Kastner, Sever, Hager, Sommer & Schmidt, 2010; Day, Campbell, Clelland & Cichowicz, 2011; Pollock, Jones, Jenkyn, Ivanova & Garland, 2012; Croft & Ribeiro, 2013; Wilson, Gissane, Gormley & Simms, 2013; Gravenhorst et al., 2014; Cuijpers, Zaal & De Poel, 2015; Gravenhorst et al., 2015b; Millar, Oldham, Hume & Renshaw, 2015; Bai, 2016).

Factors affecting rowing performance

Modelling in rowing is based on sport-scientific knowledge and the different boat classes' limitations. Typically, methods of functional diagnostics of the human motor system are derived from sports biomechanics.

The following main factors affecting the outcome of sports competitions are distinguished: technique of movement, physical fitness and tactics. This differentiation led to a methodology relying on separate assessment of physical fitness by measuring it as well as evaluating and optimizing movement techniques using mathematical modelling (Fidelus, 1970). In the vast majority of sport disciplines, the outcome of the competition depends on the training of the three main competencies:

- athlete's physical features,
- motion technique, and
- tactics (Figure 3.1).

In addition to the impact of the athlete's condition on the result of the race, the following environmental conditions are also of paramount importance in rowing:

- water current,
- wind, and
- surface conditions (Figure 3.1).

Physical features should be understood as measurable competitor attributes. The most important ones among these physical skills are the power and strength used to achieve the best possible results during a competition. Physical features are measured on special apparatus that aim to eliminate any influence of technique or environmental factors on the test results. Muscle torques in the main joints under static or isokinetic conditions, measurements on cycle ergometers (Hofmann et al., 2007), handcycles and special rowing ergometers (Torres-Moreno et al., 2000) are the most commonly used measurements. The athlete's endurance can be determined during measurements on cycle ergometers and rowing ergometers in short-term efforts. The results of power and strength measurements in the laboratory facilitate assessment of the athlete's strength potential.

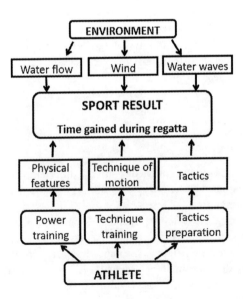

Figure 3.1 Main training factors and environmental conditions affecting the result in rowing.

Motion technique is a way of performing a movement task and strongly affects the sport result. Biomechanical methods (e.g., kinemetry, dynamometry, electromyography) are utilized to analyze motion technique.

The third most important factor that significantly affects the sports result is tactics. The term "tactics" has been known since ancient times. This concept was most often referred to a military context; tactics have represented the art of fighting. Today, tactics can be seen in all sorts of human rivalries such as politics, business and sport. In the field of sporting rivalry, tactics are particularly important in team sports and all disciplines where opponents' behavior and changing conditions of competition are unpredictable. Whenever we deal with unpredictable conditions of competition and changes in opponents' tactics, tactical assumptions and current tactical changes should result from decisions taken on the basis of probability theory and even game theory. Movement technique and tactics of sport combat are specific to each sport, with different movement techniques for each separate competition and even for each physical exercise. This specificity of technique and tactics determine the motion task that has to be solved as effectively as possible (Fidelus, 1970).

The problem of proper motor and mental preparation of athletes for competition is similar in all sports. The main goal of motor training and mental preparation is to get the greatest potential of power and mental efficiency for the competitor. Particular motor and mental characteristics development levels vary, for example in strength and sports endurance, but these are only quantitative differences. A certain minimum of all the motor and mental qualities is needed for each competitor. Tactics are inextricably linked to technique and movement coordination because the motion is precisely defined on that basis. Thus, the tactic of sport combat is a meaningful program of the specific action content, interaction and counteraction in any sports match. Optimizing motion techniques requires building a proper model of motion and behavior in accordance with the "dynamic study" procedures proposed by Cannon (1967) or Maryniak (1975) for a particular athlete. Finding the best tactical solutions also requires developing appropriate mathematical models and using them to solve probability and game theory questions. Biomechanics do not deal with the athletes' mental health, assuming they are motivated to the maximum and in the best possible state due to the best possible sports result and mental status.

In the case of rowing, as in many other sports, tactics means a conscious change of movement technique that, furthermore, means changing the paddling method. For mathematical modelling, the oar work can be characterized using the following parameters:

- paddling frequency [Hz],
- time of the active phase of the paddle's blade [s],
- time of the passive phase of the paddle's blade [s],
- time of blade dipping [s], and
- the ascent time of the paddle [s].

The active phase of the paddle begins at the moment called "the catch" when the blades are placed in the water. Full cycle time is equal to the sum of the active and passive time phases. Before the end of the stroke, the paddle handle is pushed down to lift the blade from the water; this is called "washing out." Dipping and ascending of the blade is realized during the active phase of the blade.

It seems that it is sufficient to divide the regatta time into the following four phases when one plans the tactics for rowing regattas:

- prestart,
- start,
- distance, and
- finish.

The pre-regatta tactics and technique parameters can be described by specifying the numbers of cycles or by the duration time for each of the four regatta phases. All the training factors and environmental conditions, shown in Figure 3.1, significantly affect the rowing technique. Athletes and coaches adjust their equipment appropriately; they change the technique and they set tactics to achieve the shortest time during the regatta.

In sport, the technique of motion like any human action can be evaluated by applying any criterion. Most often, it is an assessment referring to the aesthetic impressions of the observer. The main creator of the theory of human and animal motor control, Nikolai Aleksandrovich Bernstein (1896–1966), claimed that well-trained movements are smooth and deft (Bernstein, 1967). Such a rowing technique assessment method is commonly used by the coaches and depends primarily on their experience. Since filming rowing during the competition is very difficult, in practice, filming during training is used to analyze athletes' movement technique and slow-motion replay is popular. Acceleration and dynamometric methods are also used to measure the acceleration of the boat, the athletes' seats and the forces developed with the paddle (Croft & Ribeiro, 2013). During any training, the use of GPS is very popular to measure parameters online and to store position and velocity data about the boat. In addition, the slide seat motion, cadence of paddling and the physiological parameters such as heart rate are measured. The assessment, effective improvement and optimization of rowing techniques in the longer term are only possible if one employs a reliable mathematical model giving the opportunity to predict the regatta results, determined by the time to cover a certain distance.

Nowadays, one of the greatest challenges of modern sports biomechanics is still the improvement and optimization of motion technique. Fundamental work taking a bio-cybernetic approach to human movement optimization has been undertaken in particular by Herbert Hatze and by others (Hatze, 1983, 1984, 1985, 2002; Cuijpers et al., 2015).

However, there is limited work related to optimizing sport results. A few papers on motion technique optimization to achieve the best possible sports results have been suggested by Maroński et al. (Maroński, 1990, 1991, 1994,

1996; Maroński & Rogowski, 2011; Maroński & Samoraj, 2015). The papers written by Maroński et al. are innovatory due to the application of advanced optimization methods in cases of various human locomotions during sports competitions. It often appears that the description of the tested sports competition is oversimplified, resulting in too limited mathematical models. Hence, the solution of the optimization problem is too general and not really relevant for any application in real conditions during sports rivalry.

Many authors have attempted to improve and optimize selected elements of rowing technique. Important and innovative approaches to this problem have been presented by different authors (Sanderson & Martindale, 1986; Aggarwal, Murgai & Fujita, 1997; Cabrera, Ruina & Kleshnev, 2006; Kastner et al., 2010; Zhang, 2011; Ignagni, 2012; Pettersson, Nordmark & Eriksson, 2014).

Problem

In what follows, we will focus on models developed to analyze and improve rowing technique. We demonstrate the applicability of mathematical models derived from physical systems as well as data-driven models. Both approaches depend on the identification of relevant parameters and/or features controlling and/or characterizing the rowing motion. Mathematical models are valuable tools for simulating and optimizing different rowing or sculling techniques. They may be applied for forecasting regatta times and gaining insights to the influence of different parameters on the efficiency of sculling. Data-driven models are helpful for categorizing rowing techniques, for identifying motion patterns and for recognizing parameters and features suitable for predicting performance development.

Concepts and methods of modelling

Deterministic mathematical models for studying physical systems

The knowledge of mathematical modelling methodology is very useful to analyze human movement. According to Albert Einstein (1879–1955), "What we call physics comprises that group of natural sciences which base their concepts on measurements; and whose concepts and propositions lend themselves to mathematical formulation." In this way, Albert Einstein defined the most important methods of the natural sciences: measurement and mathematical modelling. Metrology is the science and practice of measurement. Metrology is defined by the International Bureau of Weights and Measures as "the science of measurement, embracing both experimental and theoretical determinations at any level of uncertainty in any field of science and technology." The ontology and international vocabulary of metrology is maintained by the International Organization for Standardization. Uncertainty is a property of a measurement result that defines the range of probable measures and values. The total uncertainty may consist of the components that are evaluated by the statistical probability of experimental data distribution. It is obvious that each measurement is subject to an error. Therefore, regarding accuracy, it is absolutely essential to present measurement results in the form of the maximum relative error, for example. Galileo Galilei (1564–1642) formulated the main idea of metrology: "Count what is countable, measure what is measurable, and what is not measurable, make measurable." The importance of mathematical modelling in science was expressed by the eminent physicist Lord Kelvin (1824–1907) who stated: "I can never satisfy myself until I can make a mechanical model of a thing. If I can make a mechanical model, I can understand it." Another eminent physicist, John von Neumann (1903–1957), defined the goal of science: "The sciences do not try to explain, they hardly even try to interpret, they mainly make models." Both measurement and mathematical modelling are independent of each other. Always while measuring, the elements of the mathematical model are considered to define the measured characteristics accurately. Similarly, it is necessary to use identification procedures consisting primarily of measurements when one creates a mathematical description of a studied phenomenon, which means creating a mathematical model. The study of complex natural processes is quite frequently based on the use of measurement and mathematical modelling. When it comes to creating a mathematical model, it is imperative to circumscribe the researched phenomenon and to define the purpose of creating this model.

Cannon proposed a procedure for studying the dynamics of mechanical systems using mathematical models (Cannon, 1967). The author proposed four stages of studying the system dynamics:

- To specify the system to be studied and imagine a physical model.
- To derive a mathematical model to represent the physical model.
- To study the mathematical model's dynamic behavior by solving differential equations of motion.
- To make decisions; i.e. to choose the system's physical parameters or to augment the system, so that it will behave as desired (Cannon, 1967).

The first step in studying the dynamics of a system is to devise an appropriate physical model. At this stage, the structure of the studied system is assumed: the number of rigid bodies, the kinematic connection class, the external forces acting on the system. At this stage of model construction, the system where the examined object is moving will be described. When considering the influence of the wind system and water current movement, as may be necessary in water sports, the aero- and hydrodynamic system should also be defined. It is also desirable to specify the names of the kinematic parameters of the object movement that will be investigated during the model experiments. For floating objects, linear velocities are defined according to the axes of the adopted reference system. Angular velocities of a sport rowing boat are determined identically to aviation and navy as rotations with respect to the axes 0x, 0y and 0z passing through the center of the boat's mass. 0x, a longitudinal axis or roll axis, is a horizontal line running through the length of the boat. 0y, a lateral or transverse axis, is a line running horizontally across the boat. 0z, an axis, or yaw axis, is a line running vertically through the boat. Traditionally, in navigation, the angular positions and angular velocities of vessels are named as follows:

- roll – is the tilting rotation of a boat about its longitudinal axis (0x),
- pitch – is the up and down rotation of a boat about its lateral axis (0y), and
- yaw – is the turning rotation of a boat about its vertical axis (0z).

In the case of modelling a sport boat rowing technique, the following forces acting on the athlete-boat-oars (ABO) system should be taken into consideration:

- gravity force;
- hydrodynamic drag: wave, viscosity, shape and volume;
- aerodynamic drag;
- buoying force; and
- oar driving force.

It is very difficult to decide on the number of degrees of freedom of the ABO system tested and adopted as the control quantities. The main problem is the difficulty in measuring or identifying a large number of parameters and functions depending on the number of degrees of freedom, which is necessary to perform the model experiments and possibly to optimize the sculling technique on a single-scull boat.

In the second stage of the sculling study, the dynamics of the system are described. The mathematical model is created in the form of a system of motion equations. There are two main approaches to creating a mathematical model of a complex process such as sculling. One is to derive the equations of motion from analytical mechanics methods using known models describing the aero- and the hydrodynamic forces (Maryniak, 1975; Wychowański, 1994; Cabrera et al., 2006). The alternative approach is to use computational fluid mechanics (CFD) and a simplified description of the dynamics of the paddler-boat-paddle system (Formaggia et al., 2008; Formaggia, Miglio, Mola & Montano, 2009; Formaggia, Mola, Parolini & Pischiutts, 2010; Sliasas & Tullis, 2010; Mattes, Schaffert, Manzer & Bohmert, 2015).

The third stage in studying the dynamics of the ABO system is to use a mathematical model to carry out computer experiments. The model experiment consists of the motion equations' integration and the calculation of the time to cover a certain distance depending on parameters describing the rowing technique and the weather conditions.

The fourth stage of the ABO study is to verify the results obtained during the simulation and to use the information obtained during the experiments to improve the paddle movement technique. In case of unsatisfactory verification of the results from the computer simulation, a change of model parameters or reconstruction of the physical model is required. Then, the whole procedure

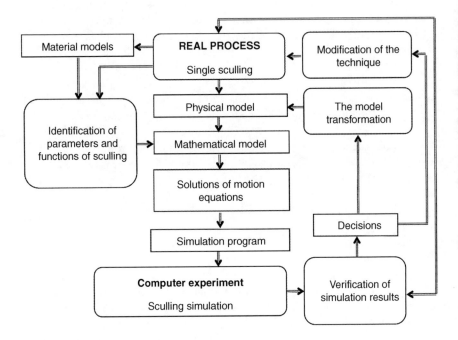

Figure 3.2 Stages of studying the dynamics of the system.

must be repeated, based on the modified model. The detailed algorithm used to study the dynamics of paddled objects is shown in Figure 3.2.

A mathematical model of the paddle technique is indispensable for its optimization. There are reports from attempts to optimize the paddle technique (Sanderson & Martindale, 1986; Aggarwal et al., 1997; Mandic, Quinney & Bell, 2004; Caplan & Gardner, 2007; Kastner et al., 2010; Ignagni, 2012; Pettersson et al., 2014; Mattes et al., 2015; Cuijpers et al., 2015) but, currently, there are no satisfactory examples for a totally effective rowing optimization.

Data-driven modelling

Complementary and as alternatives to physics-based models, data-driven models are becoming increasingly popular for analyzing sport motions and for classifying the types of motions and the types of athletes. These tendencies can also be observed in the science literature dedicated to rowing (e.g., Tachibana, Furuhashi, Shimoda, Kawakami & Fukunaga, 1999; Rauter et al., 2011; Gravenhorst, Muaremi, Draper, Galloway & Tröster, 2015a). Data-driven modelling techniques make use of computational intelligence and machine-learning methods. Among them, neural networks, Fuzzy rule-based systems, genetic algorithms and classification methods such as Support Vector Machines (SVM) or Bayesian classification have gained wide popularity (cf. Baca, 2012, 2014). The overall procedure is depicted in Figure 3.3.

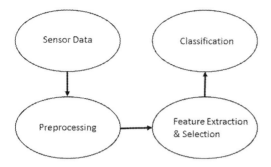

Figure 3.3 Statistical classification from sensor data.

Numerous sensors and systems are applicable in human motion analysis. They are capable of capturing a large amount of data. Typically, some preprocessing of these data is required in order to smooth the data, to remove outliers or to correct for offsets. In the next step, the often highly complex data set has to be reduced. To do so, feature extraction is accomplished, and there are different ways to do this. The easiest approach is to calculate simple statistics such as minimum, maximum, variance, average values, etc. Features may also be obtained from the application of filtering techniques, such as Fast Fourier Transform. Linear transformation techniques provide not only the means to reduce the dimensionality of the data set, but also to gain features characterizing the original data set. Among these, principal component analysis (PCA; see the paper of O'Halloran and Anderson (2008a) for an example from rowing), independent component analysis and linear discrimination analysis are quite frequently used.

In the next step, feature selection methods are applied in order to identify a subset of features from the original set which is well suited for a subsequent classification. Various optimization criteria have been established for this process. Sequential forward selection (beginning with an empty set and repeatedly adding the feature best fulfilling the optimization criterion with those already selected) and backward selection (removing features repeatedly from the set) have gained wide popularity. Finally, statistical classifiers are applied for categorizing the data. There are many powerful tools to do so. Classification trees such as Random Forest, decision engines, Bayes classifiers, k-Nearest Neighbors (k-NN), rule-based approaches, linear discriminant classifiers, Fuzzy logic techniques or SVMs may be utilized. Artificial neural networks (ANNs) may also be regarded as statistical classifiers. In particular, much attention has been given to unsupervised neural nets such as self-organizing maps (SOM) in recent years. Their application has proven to be helpful in identifying the patterns in complex motor tasks or processes. SOMs facilitate mapping of motion processes to a two-dimensional trajectory described by a sequence of neurons. Each neuron represents a unique state of this process (Baca, 2012, 2014). An example related to rowing can be found in the papers by O'Halloran & Anderson (2008b) or Perl & Baca (2003). Hidden Markov

Models (HMMs) provide another alternative for classification purposes. They are appropriate for being used in categorizing human motion and activity. Generally speaking, an HMM is a statistical model for describing the characteristics of a stochastic process. A finite number of states may be taken during a process. Each state is associated with a transition probability to other states. The state at a specific time is directly and solely influenced by the state at the previous time. After each transition from one state to the next, an output observation is generated based on an observation probability distribution. From the observable output, the underlying process may be deduced. If, for instance, different motion techniques are classified, separate HMMs will be trained for each individual motion technique. When evaluating a given sequence, likelihoods from each trained HMM are calculated. The sequence can then be assigned to the HMM with the greatest likelihood.

Methodological experience

When developing mathematical models for the rowing motion, different aspects have to be considered. Does the model only refer to the motion of the ABO system and the forces acting on it, or does it also consider the motion of the rower or team of rowers? In the latter case, biomechanical aspects of the rower(s) have to be accounted for and the coupling of the rower-boat-oars system has to be modeled adequately. Biomechanical models of the rower could be set up comprising a set of body segments (e.g., trunk, head, upper extremities, thighs and shanks), but also comprising complex (neuro-)musculoskeletal elements. These could then, for example, incorporate properties of the individual rower's muscles, which may alter because of training, making the modelling process even more complicated. However, the development of adequate models for a pure boat-oar system driven by external forces is not an easy task, either. This is particularly because of the hydrodynamic force development, which physics finds difficult to comprehend. Nevertheless, since the models built have been well suited for simulations, they may be validated by measurements.

Data-driven models partly rely on the data accuracy upon which they are based. This is particularly the case when used in the studies related to elite sports, where differences of the parameter values between top athletes may be very small. In the case of rowing, one specific problem arises from the measuring environment. Any movements in water are difficult to accomplish. There are no sensors available that provide adequate information on hydrodynamic forces on the oars and on the boat. On the other hand, a large amount of data may be captured facilitating calculation of numerous features, which may be suited for subsequent analyses.

Results

Simple mathematical model of single sculling

As a result of a mathematical model based on physical principles, equations of motion may be obtained, which then may be used to simulate motion. Thereby,

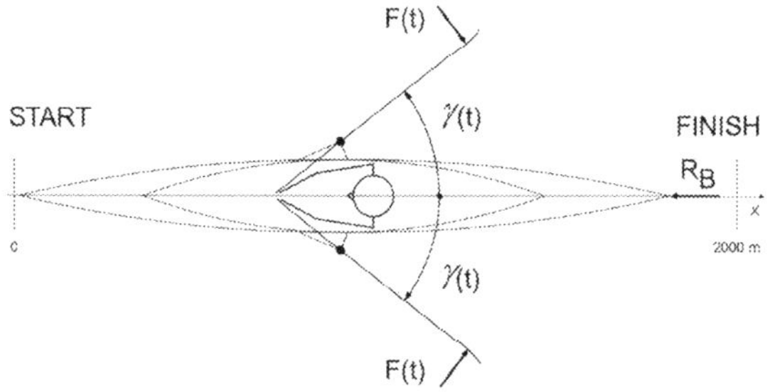

Figure 3.4 Physical model of the athlete-boat-oars system. F(t) denotes the scull force, R_B the boat's drag.

the sensitivity and impact of different parameters on the simulation result may be studied. As an example, a model for the single-scull technique shall be outlined. Single sculling is rowing with two oars, each in one hand. This discipline involves covering a specific distance in the shortest possible time and, therefore, efficient technique is of paramount importance. According to Zatsiorsky and Yakunin (1991): "Despite relatively numerous studies, the biomechanics of rowing remains poorly understood." These authors were among the first ones who attempted the mathematical description of rowing. A lot of authors have dealt with similar subjects and, among others, Pulman (2017) gives a thorough description of the phenomena concerning different aspects of rowing. Some authors, such as Kinoshita, Miyashita, Kobayashi and Hino (2008), Findlay and Turnock (2010), and Millar et al. (2015) have treated similar problems, especially the hydrodynamics phenomena related to flow past the oar. We present a mathematical model (Wychowański, 2017a) that predicts the result of a single-scull regatta. The mathematical model describes the dynamics of the ABO system.

For the physical model of rowing shown in Figure 3.4, the mathematical model for technique was derived from Newton's second law of dynamics and rules describing hydrodynamic forces.

The following second-order differential equation of the ABO system motion has been derived by Wychowański et al. (2017a):

$$m\ddot{x}(t) = T_{OA}(t) - R_B(t) = T_{OA}(t) - k_B \dot{x}(t)^2$$

$$T_{OA}(t) = 2F(t)\sin\gamma(t)$$

where:
  m – mass of the ABO system,
  $\ddot{x}(t)$ – acceleration of the boat in forward direction,
  $\dot{x}(t)$ – velocity of the boat in forward direction,
  k_B – drag coefficient,
  $R_B(t)$ – boat drag,

$F(t)$ – force generated by the oar, and
T_{OA} – propulsion force, where the hydrodynamic force $F(t)$ follows a simplified Bollay's model (1939) as follows:

$$F(t) = 0.25 c_{DOA} \rho_{H_2O} \left[1 + sgn\, \dot{\gamma}_{OA}(t)\right] \eta(t)\, sgng(t) \int_{l_{OAMIN}}^{l_{OAMAX}} \left[l_{OA} \dot{\gamma}_{OA}(t) - \dot{x}(t) \sin \gamma_{OA}(t)\right]^2 b_{OA} dl_{OA}$$

where:
c_{DOA} – drag coefficient of the oar blade,
ρ_{H_2O} – water density,
$\gamma_{OA}(t)$ – oar angle,
l_{OA} – oar length,
b_{OA} – oar width,
sgng(t) – braking or propelling function, and
$\eta(t)$ – oar diving control function period.

The left side of the ABO motion equation presents the product of the acceleration and the mass of system, and the right side of the equation, in accordance with Newton's second dynamics principle, is the sum of external forces acting on the system. The force developed by the oar $F(t)$ is the integral of the pressure acting over the oar blade surface during an oar stroke. The angular motion of the oar around the oarlock during the active phase is modeled as a cosinusoidal shape. As a result of solving the second-order differential equation given above, the position $x(t)$, speed $\dot{x}(t)$ and acceleration $\ddot{x}(t)$ of the boat during the regatta is obtained.

The method of deriving the equations for the power developed by the paddle and boat dynamics from the Navier-Stokes equations taking into account the influence of wind and water current is presented in the papers by Sługocki and Wychowański et al. (Sługocki, Wychowański, Orzechowski & Radomski, 2017; Wychowański et al., 2017b). The resulting equations of motion may be simulated for different sets of parameter values. Thereby, a single nonlinear differential equation has to be solved.

Data and features characterizing boat motion

A variety of sensors measuring kinematic data (GPS, accelerometers, gyroscopes, inertial measurement units, potentiometers) and force data (e.g., strain gauges) may be used to obtain rowing-specific parameter values. From the time functions of these parameters (e.g., distances, velocities, accelerations; pitch, yaw and roll angles of the boat; oar angles; forces applied to the oar) numerous features may be determined. Typical features obtained comprise stroke rate, the distance during one stroke, maximum and minimum velocities, pitch, yaw and roll angle and the instants these maxima and minima occur.

Based on such features, it is possible to identify types of rowers by classifying them from boat motion characteristics. Moreover, dominant rowers in the boat may be identified, based on the assumption that their individual technique mostly influences boat motion. Another field of application could be the analysis of the effect of certain training measures.

The study by Gravenhorst et al. (2015a) may serve as an excellent example. The authors measured parameters using mobile on-board measurement systems mounted on the boat and integrated in the oarlocks. From the sensor data obtained they calculated boat-, oar- and crew-specific features in a study featuring double-scull boats. Examples of features calculated were the maximum, minimum and average boat velocities during one stroke as boat-specific features, the oar absolute angle relative to the boat at the beginning and the end of the drive phase (catch and finish positions) as oar-specific features and the differences of a rower's features compared to the rower in the stroke position as crew-specific features.

Examples

Mathematical model of single sculling

Simulations have been performed based on the model presented before (Wychowański, 2017a). Parameter values were set as follows: mass of the ABO system (rower, boat, oars) $m = 100$ kg, minimal angle of an oar stroke $\gamma_{min} = 30°$, maximal angle of an oar stroke $\gamma_{max} = 120°$, drag coefficient of the boat $k_B = 3.8$ kg/m, dimensionless drag coefficient of the oar $c_{DOA} = 1.13$, water density $\rho_{H2O} = 1,000$ kg/m³, distance between an oar lock and the oar blade tip $l_{OAmax} = 2.04$ m, distance between oar lock and oar blade root $l_{OAmin} = 1.6$ m, oar blade width $b_{OA} = 0.23$ m, total rowing cycle period (while rowing at a distance, see Introduction: phases of the regatta) $T = 2$ s, active phase of a cycle period $T_A = 0.7$ s, passive phase of cycle period $T_B = 1.3$ s. Selected results for the computer simulation when rowing a distance of 2,000 m are shown in figures 3.5 and 3.6. The equation of motion was solved with the ode45 code (Shampine & Reichelt, 1997) that is based on the explicit Runge-Kutta formula of orders 4 and 5 in a MATLAB® environment. Error tolerances of 10^{-10} for absolute and 10^{-8} for relative tolerances were set. The simulated time for traveling 2,000 m was approximately 7 minutes and 55 seconds (475 s).

When replacing the limit angles $\gamma_{min} = 30°$ and $\gamma_{max} = 120°$ by $\gamma_{min} = 45°$ and $\gamma_{max} = 135°$, the simulation of the 2,000 m regatta resulted in a time of 8 minutes and 1s (481 s).

The main criterion for the evaluation of the model was the time to cover the distance of 2,000 m obtained by simulation on the basis of the assumed frequency as well as active and passive phases of rowing. A verification of the model describing the result of rowing regattas is very difficult and practically impossible to implement due to the variability of weather conditions and difficulties in maintaining the assumed rowing technique. On the basis of a cursory observation, the predicted rowing result obtained by computer simulations of the simple mathematical rowing technique is fraught with error of not more than 10%. Figures 3.5 and 3.6 show the results of a computer simulation of the start to the regatta. During this simulation, the oar blade's immersion time was $T_{IN} = 0.1$ s and the paddle ascent time was $T_{OUT} = 0.1$ s. In the prestart phase, 1 stroke was performed ($T_A = 0.7$ s, $T_B = 0.5$ s); in the start phase, 4 strokes ($T_A = 0.5$ s, $T_B = 0.5$ s); in the distance phase, 200 strokes ($T_A = 0.7$ s, $T_B = 1.3$ s) and, during the finish, strokes were performed with a period of active oar work of $T_A = 0.6$ s

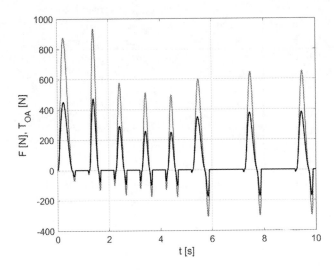

Figure 3.5 An example of single oar force F (black) and total driving force T_{OA} (gray) in the first 10 s of a regatta.

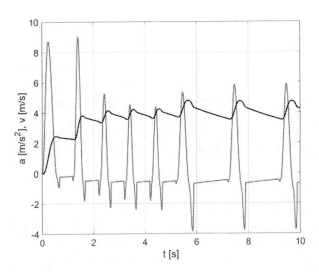

Figure 3.6 An example of boat acceleration a as italic style (gray) and boat velocity v (black) in the first 10 s of a regatta.

and passive paddle work $T_B = 1$ s. In figures 3.5 and 3.6, the frequency changes resulting from the tactics adopted above are clearly visible. Negative force values that cause boat braking arise during dipping and ascent of the paddle and are an important indicator of the quality of the rowing technique. Negative force values that cause boat braking arise during dipping and ascent of the paddle and are an important indicator of the quality of the rowing technique.

Day et al. (2011), Formaggia et al. (2008) and Formaggia, Miglio, Mola and Montano (2009) dealt with similar issues, paying attention to the hydrodynamic forces formation phenomenon on the oar. Their results describe the course of the forces on the oar blade according to the experimental data very well. As a result of their assumptions about the cosinusoidal course of an oar angle, Wychowański et al. (2017a) obtained an essentially different shape of the force variation on the blade.

The selection of different parameter values for the equation of motion enabled simulation of different rowing techniques on a single scull and a prediction of their influence on the regatta result. In more detail, the sculling model was used to analyze the impact of the movement technique on the outcome of the race over a distance of 2000 m. Simulation results of the regatta for the range of angle paddling 90° were once obtained for the limit values of the paddle angles $\gamma_{min} = 30°$ and $\gamma_{max} = 120°$ and the other time for $\gamma_{min} = 45°$ and $\gamma_{max} = 135°$. The race result for the latter angles was about 6 seconds worse. The simulation results match our experimental data and references of other authors.

Data-driven model

Again, the data-driven model presented by Gravenhorst et al. (2015a) shall serve as an example. The authors were able to construct and select features that allowed them to distinguish the technique of four rowers with high accuracy. They found out that the identified features were relevant to identifying the best-fitting rowers for crews. Moreover, boat speed could be predicted, based on a linear regression model set up from five features with a root mean square error of less than 35% of the boat velocity standard deviation.

Exercises

Following Newton's second law, the equation of the boat can be written in its simplest form as:

$$m\ddot{x}(t) = F_x(t),$$

where m is the sum of the masses of the rowers, the boat and the oars, $\ddot{x}(t)$ is the acceleration of the boat in a forward direction as a function of time and $F_x(t)$ is the sum of all external forces acting on the boat in the direction of motion.

- Reflect on the force components affecting $F_x(t)$. How can they be influenced by the rower(s)?
- Try to modify the paddle power equation and the boat motion equation (Results section) to take into account the effects of wind and water current on the paddle effect at a specified frequency.
- Suggest another mathematical description of the power driven by the paddle.
- Think of a simple ABO physical model that takes into account at least one important movement of the athlete relative to the boat.
- Consider single sculling: In addition to the boat, kinematic parameters describing the rower's motion are collected. The rower is modeled as moving symmetrically and consists of the following segments: trunk + head + upper extremities, thighs, shanks. Think of possible features suitable for differentiating between types of motion technique using data-driven models.

References

Aggarwal, R., Murgai, R. & Fujita, M. (1997). Speeding up technology-independent timing optimization by network partitioning. In Proceedings of The 1997 IEEE/ACM International Conference on Computer-Aided Design, 83–90.

Baca, A. (2012). Methods for recognition and classification of human motion patterns: A prerequisite for intelligent devices assisting in sports activities. 7th Vienna International Conference on Mathematical Modeling, February, 2012. *Mathematical Modeling*, 7(1), 55–61.

Baca, A. (2014). Adaptive systems in sport. In P. Pardalos & V. Zamaraev (Eds.), *Social networks and the economics of sports* (pp. 115–124). Cham: Springer.

Bai, G. (2016). Research on some problems on endurance training effectiveness. In Z. Zeng & X. Bai (Eds.), Proceedings of The 2016 2nd Workshop on Advanced Research and Technology in Industry Applications. Paris, France: Atlantis Press.

Bernstein, N. (1967). *The co-ordination and regulation of movements*. Oxford: Pergamon Press.

Bollay, W. (1939). A non-linear wing theory and its application to rectangular wings of small aspect ratio. *Zeitschrift für angewandte Mathematik und Mechanik*, 19, 21–35.

Cabrera, D., Ruina, A. & Kleshnev, V. (2006). A simple 1(+) dimensional model of rowing mimics observed forces and motions. *Human Movement Science*, 25, 192–220.

Cannon, R. (1967). *Dynamics of physical systems*. New York: McGraw-Hill.

Caplan, N. & Gardner, T. (2007). Optimization of oar blade design for improved performance in rowing. *Journal of Sports Sciences*, 25, 1471–1478.

Croft, H. & Ribeiro, D. (2013). Developing and applying a tri-axial accelerometer sensor for measuring real time kayak cadence. *Procedia Engineering*, 60, 16–21.

Cuijpers, L., Zaal, F. & De Poel, H. (2015). Rowing crew coordination dynamics at increasing stroke rates. *PLoS One*, 10.

Day, A., Campbell, I., Clelland, D. & Cichowicz, J. (2011). An experimental study of unsteady hydrodynamics of a single scull. *Proceedings of the Institution of Mechanical Engineers, Part M. Journal of Engineering for the Maritime Environment*, 225, 282–294.

Dworak, L. (2010). Sports biomechanics in the research of the Department of Biomechanics of University School of Physical Education in Poznan. Part 2. Biomechanics of rowing: Research conducted in the rowing pool and under real conditions. Reconstruction and synthesis. *Acta of Bioengineering and Biomechanics*, 12, 103–112.

Fidelus, K. (1970). *The place and importance of movement technique in sport theory* (in Polish). Warsaw: Sport i Turystyka.

Findlay, M. & Turnock, S. (2010). Mechanics of a rowing stroke: Surge speed variations of a single scull. *Proceedings of the Institution of Mechanical Engineers Part P-Journal of Sports Engineering and Technology*, 224, 89–100.

Formaggia, L., Miglio, E., Mola, A. & Montano, A. (2008). Fluid structure interaction problems in free surface flows: Application to boat dynamics. *International Journal for Numerical Methods in Fluids*, 56, 965–978.

Formaggia, L., Miglio, E., Mola, A. & Montano, A. (2009). A model for the dynamics of rowing boats. *International Journal for Numerical Methods in Fluids*, 61, 119–143.

Formaggia, L., Mola, A., Parolini, N. & Pischiutts, M. (2010). A three-dimensional model for the dynamics and hydrodynamics of rowing boats. *Proceedings of the Institution of Mechanical Engineers Part P-Journal of Sports Engineering and Technology*, 224, 51–61.

Gravenhorst, F., Muaremi, A., Draper, C., Galloway, M. & Tröster, G. (2015a). Identifying unique biomechanical fingerprints for rowers and correlations with boat speed? A data-driven approach for rowing performance analysis. *International Journal of Computer Science in Sport*, 14(1), 4–33.

Gravenhorst, F., Thiem, C., Tessendorf, B., Adelsberger, R., Arnrich, B., Draper, C., ... Troster, G. (2015b). SonicSeat: Design and evaluation of a seat position tracker based on ultrasonic sound measurements for rowing technique analysis. *Journal of Ambient Intelligence and Humanized Computing*, 6, 613–622.

Gravenhorst, F., Muaremi, A., Kottmann, F., Tröster, G., Sigrist, R., Gerig, N. & Draper, C. (2014). Strap and row: Rowing technique analysis based on inertial measurement units implemented in mobile phones. IEEE 9th International Conference on Intelligent Sensors, Sensor Networks and Information Processing (IEEE ISSNIP 2014) (pp. 1–6). April 21–24, 2014, Singapore. Piscataway, NJ: IEEE.

Hatze H. (1981). *Myocybernetic control models of skeletal muscle.* Pretoria: University of South Africa Press.

Hatze, H. (1974). The meaning of the term 'biomechanics'. *Journal of Biomechanics*, 7(2), 189–190.

Hatze, H. (1983). Computerized optimization of sports motions: An overview of possibilities, methods and recent developments. *Journal of Sports Sciences*, 1, 3–12.

Hatze, H. (1984). Quantitative analysis, synthesis and optimization of human motion. *Human Movement Science*, 3, 5–25.

Hatze, H. (1985). Dynamics of the musculoskeletal system. *Journal of Biomechanics*, 18(7), 515.

Hatze, H. (2002). Fundamental issues, recent advances, and future directions in myodynamics. *Journal of Electromyography and Kinesiology*, 12, 447–454.

Hofmann, P., Jurimae, T., Jurimae, J., Purge, P., Maestu, J., Wonisch, M., ... von Duvillard, S. (2007). HRTP, prolonged ergometer exercise, and single sculling. *International Journal of Sports Medicine*, 28, 964–969.

Ignagni, M. (2012). Optimal sculling and coning algorithms for analog-sensor systems. *Journal of Guidance Control and Dynamics*, 35, 851–860.

Kastner, M., Sever, A., Hager, C., Sommer, T. & Schmidt, S. (2010). Smart phone application for real-time optimization of rower movements. In A. Sabo, P. Kafka, S. Litzenberger & C. Sabo (Eds.), *Engineering of sport 8: Engineering emotion.* 8th Conference of the International Sports Engineering Association.

Kinoshita, T., Miyashita, M., Kobayashi, H. & Hino, T. (2008). Rowing velocity prediction program with estimating hydrodynamic load acting on an oar blade. In N. Kato & S. Kamimura (Eds.), *Bio-mechanisms of swimming and flying* (pp. 345–359). Tokyo: Springer Japan.

Mandic, S., Quinney, H. & Bell, G. (2004). Modification of the wingate anaerobic power test for rowing: Optimization of the resistance setting. *International Journal of Sports Medicine*, 25, 409–414.
Maroński, R., (1994). On optimal velocity during cycling. *Journal of Biomechanics*, 27, 205–213.
Maroński, R. & Rogowski, K. (2011). Minimum-time running: A numerical approach. *Acta of Bioengineering and Biomechanics*, 13, 83–86.
Maroński, R. & Samoraj, P. (2015). Optimal velocity in the race over variable slope trace. *Acta of Bioengineering and Biomechanics*, 17, 149–153.
Maroński, R. (1990). On optimal running downhill on skis. *Journal of Biomechanics*, 23, 435–439.
Maroński, R. (1991). Optimal distance from the implement to the axis of rotation in hammer and discus throws. *Journal of Biomechanics*, 24, 999–1005.
Maroński, R. (1996). Minimum-time running and swimming: An optimal control approach. *Journal of Biomechanics*, 29, 245–249.
Maryniak, J. (1975). *Dynamic theory of moving objects* (in Polish). Łódź: Wydawnictwa Politechnika.
Mattes, K., Schaffert, N., Manzer, S. & Bohmert, W. (2015). Non-oarside-arm pull to increase the propulsion in sweep-oar rowing. *International Journal of Performance Analysis in Sport*, 15, 1124–1134.
Millar, S., Oldham, A., Hume, P. & Renshaw, I. (2015). Using rowers' perceptions of on-water stroke success to evaluate sculling catch efficiency variables via a boat instrumentation system. *Sports*, 3, 335–345.
Morecki, A., Ekiel, J. & Fidelus, K. (1971). *Bionics of motion* (in Polish). Warsaw: PWN.
O'Halloran, J. & Anderson, R. (2008a). Deterministic component extraction using PCA for evaluation of rowing data. In ISBS-Conference Proceedings Archive (Vol. 1, No. 1).
O'Halloran, J. & Anderson, R. (2008b). The use of Kohonen Feature Maps in the kinematic analysis of rowing performance. In Proceedings of The 26th Conference of the International Society of Biomechanics in Sport, 165–168.
Perl, J. & Baca, A. (2003). Application of neural networks to analyze performance in sports. In E. Müller, H. Schwameder, G. Zallinger & V. Fastenbauer (Eds.), Proceedings of The 8th Annual congress of the European College of Sport Science, 342.
Pettersson, R., Nordmark, A. & Eriksson, A. (2014). Simulation of rowing in an optimization context. *Multibody System Dynamics*, 32, 337–356.
Pollock, C., Jones, I., Jenkyn, T., Ivanova, T. & Garland, S. (2012). Changes in kinematics and trunk electromyography during a 2000 m race simulation in elite female rowers. *Scandinavian Journal of Medicine & Science in Sports*, 22, 478–487.
Pulman, C. (2017). *The physics of rowing*. Cambridge, UK: Ithaca. http://eodg.atm.ox.ac.uk/user/dudhia/rowing/physics/rowing.pdf
Rauter, G., Sigrist, R., Baur, K., Baumgartner, L., Riener, R. & Wolf, P. (2011). A virtual trainer concept for robot-assisted human motor learning in rowing. *BIO Web of Conferences 1*, 72.
Sanderson, B. & Martindale, W. (1986). Towards optimizing rowing technique. *Medicine and Science in Sports and Exercise*, 18, 454–468.
Shampine, L. & Reichelt, M. (1997). The MATLAB ODE Suite. *SIAM Journal on Scientific Computing*, 18(1), 1–22.
Sliasas, A. & Tullis, S. (2010). The dynamic flow behaviour of an oar blade in motion using a hydrodynamics-based shell-velocity-coupled model of a rowing stroke. *Proceedings of the Institution of Mechanical Engineers Part P-Journal of Sports Engineering and Technology*, 224, 9–24.

Sługocki, G., Wychowański, M., Orzechowski, G. & Radomski, D. (2017). The hydrodynamic flow past the oar and mathematical description of the oar force. In Ł. Stettner (Ed.), Proceedings of The 46th National Conference on the Applications of Mathematics, Siwarna in Zakopane-Kościelisko, September 5–12, 2017. Retrieved: www.impan.pl/~zakopane/46/Sługocki.pdf. Institute of Mathematics Polish Academy of Science.

Tachibana, K., Furuhashi, T., Shimoda, M., Kawakami, Y. & Fukunaga, T. (1999). An Application of Fuzzy Modeling to Rowing Motion Analysism. IEEE Conference on Systems, Man, and Cybernetics, 190–195.

Torres-Moreno, R., Tanaka, C. & Penney, K. (2000). Joint excursion, handle velocity, and applied force: A biomechanical analysis of ergonometric rowing. *International Journal of Sports Medicine*, 21, 41–44.

Wilson, F., Gissane, C., Gormley, J. & Simms, C. (2013). Sagittal plane motion of the lumbar spine during ergometer and single scull rowing. *Sports Biomechanics*, 12, 132–142.

Wychowański, M., Nosarzewski, Z. & Karpiłowski, B. (1987a). Stand for training and measuring kayak loads. Patent description No. P263645, January 14, 1987. Urząd Patentowy Rzeczpospolitej Polskiej.

Wychowański, M., Nosarzewski, Z. & Witkowski, M. (1987b). Oar with a deformable surface of the blade. Patent description No. P263646, January 14, 1987. Urząd Patentowy Rzeczpospolitej Polskiej.

Wychowański M., Pośnik J., Nosarzewski Z., Wojtaś H., Staniak Z. & Witkowski M. (1989). An approach to optimizing the length of the kayak paddle. *Biology of Sport*, 6(3), 235–240.

Wychowański, M. (1994). Mechanics of the kayak-athlete-paddle system (in Polish). Ph.D. thesis under the supervision of Jerzy Wicher, Warsaw University of Technology Faculty of Automotive and Construction Machinery Engineering. Retrieved: Main Library of the Warsaw University of Technology.

Wychowański, M., Sługocki, G., Orzechowski, G., Staniak, Z., Radomski, D., Trzciński, A. & Wit, A. (2017a). A simple mathematical model of a single sculling technique. In W. Potthast, A. Niehoff & S. David (Eds.), Proceedings of The 35th Conference of the International Society of Biomechanics in Sports German Sport University Cologne, Cologne, Germany.

Wychowański, M., Sługocki, G., Orzechowski, G., Staniak, Z., Radomski, D. & Wit, A. (2017b). Modeling of a single sculling technique. In Ł. Stettner, (Ed.), Proceedings of The 46th National Conference on the Applications of Mathematics, Siwarna in Zakopane-Kościelisko, Poland, September 5–12, 2017.

Zatsiorsky, V. & Yakunin, N. (1991). Mechanics and biomechanics of rowing: A review. *International Journal of Sport Biomechanics*, 7(3), 229–281.

Zhang, Z. (2011). The optimization study on the shelterbelts of the rowing area. In G. Lee, (Ed.), 2011 2nd International Conference on Information, Communication and Education Application (pp. 156–161), Bucharest, Romania.

Part II
Team sports and the modelling of playing processes and tactical behavior

4 Soccer
Process and interaction

Jürgen Perl and Daniel Memmert

Introduction

As far as modelling is concerned, soccer—from a tactical as well as a strategic point of view—is one of the most thrilling and challenging sports (Memmert, Lemmink & Sampaio, 2017). This is due to the fact that the number of players, the velocities of the players and the ball and the size of the pitch are in an optimal relation to each other, generating complex patterns of technical and tactical facets and components in various strategic frames (Memmert & Perl, 2009b; Frencken, Poel, Visscher & Lemmink, 2012; Gonçalves, Figueira, Maças & Sampaio, 2014). But there is more to the challenge than that complexity alone (Duarte et al., 2012). Here, action-based and process-based models are discussed to distinguish the advantages and disadvantages of analyzing actions vs. interactions in sports (Lames & McGarry, 2007; McGarry, 2009).

The basic reason for the problem of modelling soccer is the data (Rein & Memmert, 2016). Either, if recorded manually, there is too little data to describe situations or activities adequately or, if recorded automatically, there is too much data for detecting striking patterns and structures.

On the next level, if enough data are available as well as adequate selection of algorithms, the modelling concepts and methods are the problems: Modelling means reducing, in particular reducing unnecessary information. In order to decide which information is necessary, at least two aspects are of importance: What is the explicit key information about a match (e.g., number of successful attacks), and what implicitly contained information (e.g. quality of passes) is needed to generate that key information?

Normally, the main focus is on the success of a team, measured by goals, points and rankings. In order to improve those scores, causes for scoring goals (i.e., offense quality) and causes for preventing goals (defense quality) have to be modeled. There are a comparably great number of very simple models that just count numbers of specific states or events like ball possessions or passes (see Chapter 8 on KPIs). One problem of those indicators is that they normally do not take into consideration implicitly contained information (e.g., about tactical patterns or quality of passes) but waste important explicit information like spatio-temporal connections between counted states or events. From the standpoint of modelling, the models underlying those indicators reduce too drastically, therefore completely failing in mapping the dynamics of the play.

Example: In a match, if one team dominates ball possession and passes, it does not necessarily mean that the team is successful, e.g., if ball possession and passes take place in the midfield and do not generate any threat in the opponent's goal area. This is the problem of the well-known strategy of "tiki-taka," which is discussed in more detail in Chapter 8 on KPIs. In turn, the numbers of ball possessions and passes can form important information if they are imbedded in information about tactics and regarding dynamic patterns.

Conclusion: On the one hand, modelling has to reduce the complexity of the modeled system without, on the other hand, losing (too much) information about the system's dynamics.

In this chapter, we summarize the current state of the art on challenges and subjects of modelling as it relates to sports. Subsequently, concepts and methods of modelling (position data, pass analysis, Voronoi cells, neural networks and creativity) as well as methodological experience (formations, Voronoi cells and pass analysis) are introduced. Finally, the complexity of processes and tactical patterns as well as quantitative vs. qualitative information are presented. In the last two sections we consider areas of application as well as training conceptions and methods.

Project/problem

Challenge of modelling

Different from technical modelling, modelling of human interaction, as in soccer, has to distinguish between "act" and "behave," where behavior can be observed, while the intention behind acting normally stays unobservable. Dealing with the example of passes, this could mean that—depending on the technical skills of the player—a pass was meant to reach player 9 but in fact reached player 10. Obviously, if the result "pass to player 10" is discussed from the viewpoint of tactical skills of the passing player, the pass model of the play cannot represent the playing dynamics precisely.

This leads to the following questions:

- How precisely can an intentionally characterized dynamic system like a play be modeled in the first place?
- What are the consequences of modelling and analysis errors for results and evaluations?

As pointed out, it is the fundamental idea of modelling to get a simplified projection of the modeled system, and it is also clear that simplification causes errors. Therefore, it is a question of weighing up how far errors are tolerable compared to the usefulness models offer through reduction of complexity, computability of system states and events and simulation of system development. It always has to be kept in mind that a model is only a reduced image of reality that does not determine reality but has to be adjusted to it.

Basic structures a model of a game has to reflect are the syntactic and semantic rules of playing:

First of all, there is the syntactic control system of rules that configure the game, e.g., with regard to fairness, players' health and in order to assess and compare results. Those rules build a limiting model of the game by defining a space of allowed activities and their formal evaluations.

The next level of syntactic rules is derived from the logic of the game. For example, in a game that is characterized by offense-defense interaction, a team that has lost possession of the ball can only reactivate its offense activities by recovering the ball. This way, rules can be generated that describe the logic model of the game.

Both types of syntactic rules are related to the basic idea of the game and are invariant against semantic interpretations like tactics or strategies.

Different from those syntactic rules, semantic rules for tactical or strategic behavior can be developed, depending on the strengths and weaknesses of the respective teams. Those rules in the positive case optimize the playing processes. One negative aspect of semantic rules, however, is that players have to conform to orders that may have been conceived (from the players' standpoint) from abstract formulas that are incomprehensible; freedom of action on the pitch is reduced and there is a danger that playing degenerates to a kind of "pitch chess" that could also reduce the interest of the crowd.

Obviously, a main problem of modelling a complex strategic game like soccer is how to combine the syntactic and the semantic levels of modelling in such a way that observable facts from the match give useful information about the quality of playing behavior without restricting individual activities too much.

This is difficult!

In the following, a brief hierarchy of approaches is presented, introducing the main ideas from getting information to modelling behavior in (sports like) soccer.

Subjects of modelling

As pointed out, in practice, modelling of sports normally means transforming observable data into information about success (Memmert & Perl, 2009a).

The simplest way of getting (useful) information from state or event data is to count the numbers of positive or negative states or events. This is the basic idea of Key Performance Indicators (KPIs; see Chapter 8 on KPIs). The first step of modelling KPIs is to find indicators that seem to reflect success. Very simple examples are the frequency of ball possessions, the number of passes or the lengths of passes. If those indicators are really success indicators, the playing success should increase with increasing indicator values. This effect can hardly be checked during one play. Correlations between indicator values and success over a series of plays provides information about the usefulness of the model in question. However, correlation does not mean logic deduction, i.e., even if success is highly correlated to the number of passes, an increasing number of passes in a play does not necessarily improve the success in that play. For that reason, context plays an important role. In the case of passes, context could be the positions (e.g., from–to) and the time (e.g., first half–second half). Therefore, in order to access a play individually by means of KPIs, it seems to be helpful to take time- and space-depending distributions of KPIs into consideration. Moreover, playing means interacting and, therefore, the pass model should not only consider the where and when but also (and in particular) the situation in which the pass took place.

This last aspect leads from a static type of modelling to a more dynamic one, where spatio-temporal patterns play an important role. Of course, whether or not a pass as an action is successful first of all depends on the technical aspects of passing and receiving. But a single successful pass normally does not generate danger; a single pass is only one action in a sequence of actions from ball recovery to gaining space and generating pressure to creating a promising situation. This means that the success of the single pass depends on its interaction with the other passes and actions of the whole process in which it is imbedded. With the item process, modelling leaves the area of KPIs and opens the view on tactical behavior and strategies. But before dealing with those complicated dynamics, the question is how a process can be modeled. First of all, a process is a time-dependent sequence of steps of different types that, combined, also give the resulting process a specific type. One example has been given above: An attacking process starts with the first step of recovering the ball followed by some steps of passing the ball directed toward the opposing goal, interrupted perhaps by returns or duels, and finally closed by a shot on goal or losing the ball. The item success of that process model can be defined quite easily by "finally closed by a (successful) shot." Regarding the process steps, "success" is much harder to define. The recovery of the ball to start the process and the winning of duels to continue the process are doubtless successful steps. The final step of taking a shot from the tactical point of view can be successful, but if done from the wrong point or in the wrong direction can be technically unsuccessful. The same holds for single passes; they can be tactically successful but technically unsuccessful or vice versa. This problem often arises and leads to comments like "poorly played but won" or "perfectly played but lost." Obviously, the challenge for modelling and evaluation of processes is to distinguish technical and temporally limited behavior from tactical and process behavior.

Nevertheless, at first glance, it seems that the success of a process in the end strongly depends on the technical skills of the players. On one hand, this is true, as bad players are hardly able to get demanding tactical ideas to work. On the other, even the best players will fail to win a match if the tactical idea is wrong, at least against a strong opponent.

This also means that the process on the higher level of tactics has to be imbedded into concepts, where the concepts as well as their success aspects can and must be modeled, too.

One main aspect is the tactical pattern of the players, i.e., the spatial relation of the players of a tactical group (e.g., defense, offense) to each other. The defense group, for instance, can build a four-player chain, a rhombus or different geometric patterns depending on the tactical situation and the offense pattern of the attacking team (this example will be discussed in more detail below on p. 79 and p. 88). The question therefore arises which defense pattern type is most successful against a specific offense pattern type. This approach is somewhat similar to chess, where only the beginner will try to capture a figure quickly while the advanced player tries to develop positional superiority. And indeed, in extreme forms of determining the playing process by rigid patterns (e.g., the tiki-taka concept), the playing dynamics become dramatically reduced in favor of waiting for and preparing chances. Nevertheless, in playing against a strong team, tactical concepts are absolutely necessary to enforce the strengths of one's own players, to reduce the strengths of the opposition players and, last but not least, to avoid running into tactical traps.

On this high and rather abstract level of planning, much like chess, simulation and prediction of systems' behavior become important aspects of modelling.

If, for example, coach A (of team A) assumes that coach B (of team B) chooses the tactical variant VB2, they must predict which of their tactical variants VA1–VA6 would be most successful against VB2. Such a prediction needs information and experience and can be complex and difficult—in particular when you consider that coach B might anticipate coach A's tactic and change their variant. This problem of dynamic adjusting of strategies is dealt with in mathematical game theory, which is not normally a field in which coaches have experience. However, appropriate models that allow for simulation can help to analyze such complicated situations and generate useful information and solutions that an experienced IT co-trainer could pass on and explain to the coach.

In the case of tactical variants, a first step of improvement could be a simple matrix containing the success values of the pairs of variants that would help the coach to decide when and how to change the tactical patterns (see p. 80).

Concepts and methods of modelling

Position data

Modelling dynamic systems like soccer plays needs data that describe the dynamics (Fonseca, Milho, Travassos & Araújo, 2012; Vilar, Araújo, Davids & Bar-Yam, 2013; Carling, Wright, Nelson & Bradley, 2014) and procedures that derive useful information from that data. For years, match analysis could only be conducted on the basis of theory because of a lack of data. Only exemplary analyses were possible, where data had to be recorded manually with huge effort online from the match or offline from a video recorder. In the past decade, however, automatically recorded position data of the players has become available, opening new dimensions in analyzing and modelling events, situations and dynamic processes in plays.

However, data are just numbers and do not necessarily represent useful information. If the aim is simply to reproduce the recorded play in the form of computer animation, then the position data itself is helpful. If the aim is to identify and evaluate passes, however, then clever algorithms are necessary that transform position data into information about events, situations and processes.

Obviously, not all interesting information about events, situations and processes can be derived from position data. Relatively easy approaches involve physical models that calculate speed, running directions or running distances of players. Much more difficult to model is the coincidence of player and ball, which is the basic information for ball control and passes up to complex tactical processes. Therefore, modelling basically has to take into consideration:

- What data is available?
- What information is interesting?
- What procedures are necessary for that data information transfer?
- Are the results accurate enough to answer the analysis questions?

The following parts will introduce some approaches dealing with positional modelling of events, situations and processes in soccer.

Neural networks

There are mainly two types of artificial neural networks in use: feed forward or backpropagation networks are used for multidimensional function calculation (see Perl, 2015), while self-organizing maps (SOMs) are used for pattern recognition (see Grunz, Perl & Memmert, 2012). As explained above, pattern recognition can play an important role in recognizing and developing tactical behavior. Therefore, in the following, the functionality of SOMs is briefly explained.

A SOM consists of neurons that contain information in the form of attribute vectors. After initializing the neurons with start vectors, through an iterative process, the network is trained using training vectors as follows:

Inputting a training vector to the network causes the identification of the winner neuron, i.e., the neuron whose attribute vector is most similar to that of the input vector (Perl & Memmert, 2010). The attribute vector of the winner neuron with a certain learning rate is adapted to the input vector. Additionally, the neighbored neurons are affected respectively, with decreasing learning rates at increasing distances to the winner neuron. As a result of this explorative learning, where neighbored neurons tend to receive similar information, neurons form clusters based on their similar input.

The most representative neuron of such a cluster can be taken as one prototype of the learned attribute vectors or patterns, i.e., after the training, the network offers the representative prototypes of the learned patterns.

If, after training, an attribute vector or pattern is inputted to the trained network, the best-fitting winner neuron is identified, the associated cluster of which then represents the (proto-)type the input pattern belongs to.

Doing it this way, a great number of patterns can be classified and replaced by a small number of prototypes. Such a procedure is particularly helpful in analyzing processes that are described by time-dependent changing states, such as the supply of material in a production process, the distribution of cars in a traffic process or the positions of players in a playing process. All those situations can be encoded by attribute vectors, meaning volumes, numbers or position coordinates, respectively. By means of a properly trained network, the set of attribute vectors is replaced by a set of prototypes that allows for a relatively easy description and characterization of process steps and process types (Perl & Memmert, 2011; Perl, Grunz & Memmert, 2013; Perl, 2015). More information about the network approach is given in Chapter 8 on KPIs.

In team games like soccer, handball, basketball or hockey, the positions of the players play an important role because they enable as well as characterize tactical ideas. In soccer, the defense players and the offense players form tactical groups, respectively, the members of which at each point in time form geometric patterns that can be described by attribute vectors, where the attributes encode the position coordinates.

Figure 4.1 shows a green board upon which the coach is describing a tactical idea using positions, position groups and resulting processes.

Recording position groups each second per half of a soccer match would result in 2700 patterns per half, which is obviously too much for useful analyses. Even a neural network could not help because the patterns, even if geometrically similar, can appear at quite different positions of the pitch and therefore cannot be identified as members of patterns.

However, if abstracted, the geometric formations—the geometric relations between player positions—can easily be decoded and trained to a network

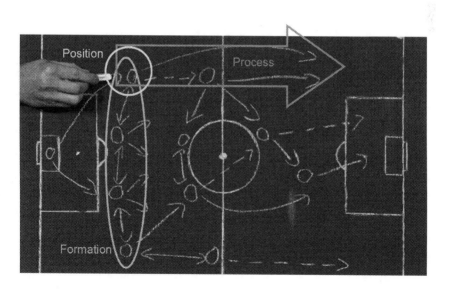

Figure 4.1 Position, formation and process on a coach's black board.

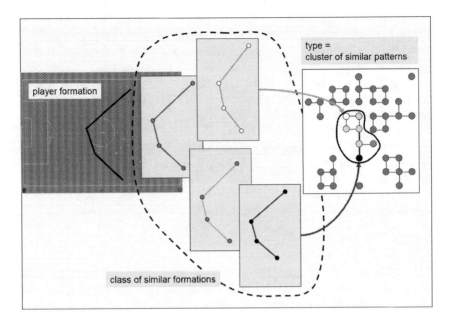

Figure 4.2 The geometric relations between player positions build a formation of the players, the similarity of which can be recognized by means of a respectively trained network.

(see Figure 4.2) that helps to identify about 10 to 30 characteristic formation types, as shown in Figure 4.3.

Once the formation types have been specified, they can be used for several kinds of process description and analysis. One such example is demonstrated in Figure 4.4, showing the matrix of coincidences between defense formation types of team A against offense formation types of team B. Each cell of the matrix contains the regarding number of coincidences.

Moreover, after having defined "success" as a measurable attribute of a state or event—e.g., ball control—the success of each pair of types can be measured regarding team A or team B, as shown in Figure 4.5.

Finally, the success matrix can be used for a first and rather simple kind of simulation of improved tactics: As Figure 4.5 shows, only 14% of defense type 2 (black marks) was successful against offense type 3. The column of possible defense types shows, however, that defense type 4 (gray marks) would perform much better against offense type 3. In turn, team B could look for its best formation types, i.e., those types the possible corresponding types of team A have success values below 50%. The strategic concept of team B then could be to avoid all non-optimal formations and, analogously, team A could do the same. The result of such a "strategic analysis" could be as shown in Figure 4.6.

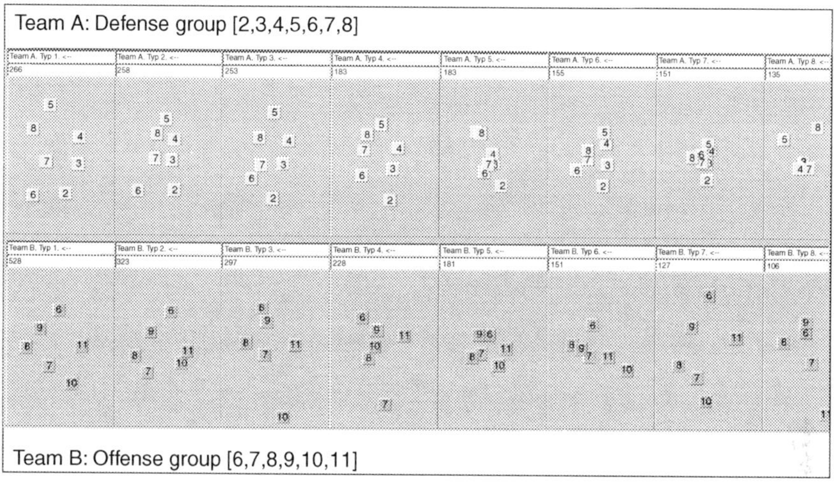

Figure 4.3 Examples of defense and offense formation types together in the order of their frequencies.

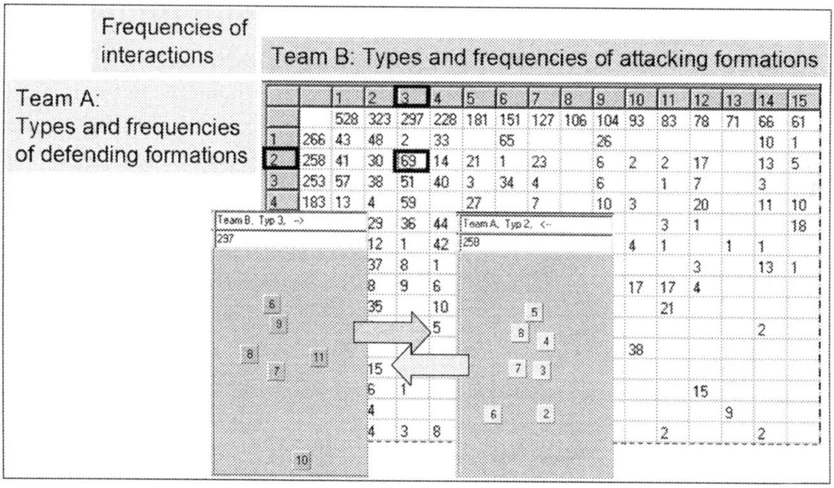

Figure 4.4 Coincidence matrix of defense formation types of team A (vertical order) against offense formation types of team B (horizontal order). The overlaid graphic shows the patterns of the selected pair of types (marked matrix element).

The same procedure could be repeated in the next and final step, leading to the result that team B always attacks in formation 12, while team A always reacts with formation 3. Obviously, this is a nonsense scenario; not only would it destroy the playing dynamics (which would be of secondary interest if it helped to

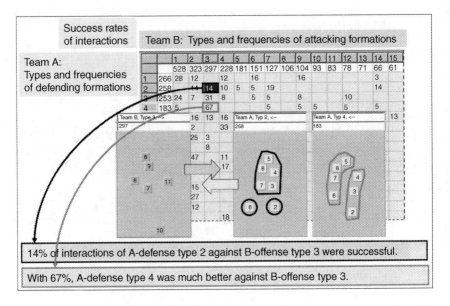

Figure 4.5 Success matrix of the coincidence matrix from Figure 4.4. The overlaid graphic shows two types of defense formation against an offense formation in order to optimize defense success.

B\A	4	7	8	9	11	12	13	14	15
1				16					
2								14	
3					10				
5									13
6	33								
8				38					
14			25			12			
15		35							

Figure 4.6 Reduced success matrix containing only successful interactions of team B. The highlighted element shows the interaction that is best for team B.

win the play), but also it simply would not happen because team A would look for alternatives to surprise team B and so, from bottom to top, step by step, team A would develop a new system of formations and interactions. This is what is analyzed systematically in mathematical game theory with the result that, normally, the best strategy is a mix of different sub-strategies. Not all of them can necessarily be optimal, but the mix means that the opposing team is always faced with uncertainty about what will happen next. Therefore, surprising actions and creativity play an important role in playing, as is pointed out later on p. 87.

So far, the focus of formation analysis has been on the temporal tactical interaction of acting and reacting formations. Another and more strategic kind of formation analysis deals with formation sequences or processes. For example, assume that team B starts its attack with formation B3, to which team A responds with A5. In turn, team B adopts B7, to which team A responds by adopting A1, and so on. The result is a pair of formation sequences of B and of A, until B's attack is finished. An alternative pair of sequences would be created if, in the second step, B adopted B2, which could result in different situations with different chances of success. In the end, there is a set of pairs of sequences with different success rates the distribution and frequencies of which give a lot of information about the strategic concepts (or at least about the behavior) of the respective teams.

Voronoi cells

One important aspect of successful attacking is how much space the attacking team controls in the critical areas of the defending team.

Before dealing with the modelling of space control in more detail (see Chapter 8 on KPIs), two central terms have to be explained:

"Critical area" can be any area in front of the opponent's goal where offensive actions can generate danger, e.g., the penalty area and the 30 m area. In Figure 4.7, the 30 m areas are marked with gray lines, while the penalty areas are marked by black boxes.

"Controlled space" of a player at a point in time is the set of all points on the pitch the player can reach faster than any other player. This set of points is called the Voronoi cell of the player at time t (Fonseca et al., 2012; Taki & Hasegawa, 2000; Kim, 2004). Voronoi cells have been used for clustering on artificial neural networks of type SOM (see p. 78), for instance, where the cluster is the set of all neurons that, regarding the contained information, are most similar or closest to a representing prototype neuron.

The difference between neurons and players, however, is that neurons do not move while players move with different accelerations and velocities and with different orientation in different directions. Therefore, the question is whether Voronoi cells can be used for modelling dynamic behavior. Of course, if the aim is to map the running behavior of players in detail, e.g., in order to simulate possible improvements, modelling has to be as precise as possible, including all information mentioned. But if the focus lies not on this individual player information but on the entire team's tactical performance, an idealization of speed and optimal orientation of individuals is barely relevant to the question. Instead, the combination of the Voronoi cells of all relevant players of the attacking team, i.e., the space controlled by the attacking team at t, gives important information about the team's offensive effectiveness at time t.

Space control can be measured as the percentage rate of the critical areas over a given time grid. Figure 4.8 shows an example of attacking team B, using a time

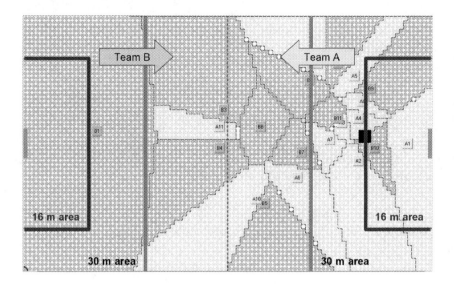

Figure 4.7 Voronoi cells of the players of team A (light gray) and team B (dark gray) on a pitch with marked main attacking areas. The position of the ball is marked by a black square.

grid of seconds. The black profile shows the percentage rates of space control in the 16 m area, while the gray profile shows the rates in the 30 m area.

Two aspects are of interest: On the one hand, one can easily recognize the attacking dynamics like phases, frequencies, durations or pressure of team B's attacks. On the other hand, however, this information is not that useful unless combined with additional information like ball possession in, or passes into, these critical areas during phases of high space control. This is the reason why, in what follows, a model of pass analysis is introduced, where the number of passes is not the main focus but rather those passes in the context of the overall playing process.

Pass analysis

Passes can be interesting and analyzed under several aspects (Memmert et al., 2017). The easiest but least informative way is to just count the number of passes. Additional information could be the direction of passes, e.g., across, forward and backward or, more precisely, the angle of the passes relative to a given coordinate system. Also, the number of outplayed opposing players could be of interest, as shown in Figure 4.9.

An important tactical context could be the formations of the team doing the passing. One example is the process of starting an attack after recovering the ball. Here, the pair of formations of the defending and passing group and the receiving and attacking group can be of interest. A second example is the progress of the

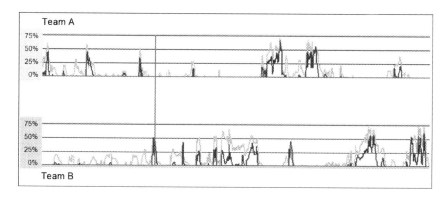

Figure 4.8 Space-control profiles of team A and team B. Black profiles show the control rates in the opponent's 16 m area, the gray profiles show the control rates in the opponent's 30 m area. The vertical gray line marks the point in time of the situation from Figure 4.7 with a control rate of almost 50% of team B in the 16 m area.

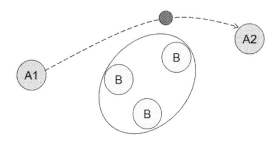

Figure 4.9 A pass from player A1 to player A2, passing three opposition players.

attacking process. Here, the change of the attacking formation from passing to receiving can be of interest.

Another important tactical context of a pass, in particular in the case of attacking play, is space control, i.e., whether a pass into the opponent's area coincides with or is followed by a high rate of space control.

While the approach of counting numbers provides only little information about tactical behavior, the approaches dealing with formations and space control provide a lot of tactical information, but are not so easy to record or to handle. One interesting approach that lies somewhere in between, providing quantitative as well as qualitative information, is the model of Pass Efficiency (see Rein, Raabe & Memmert, 2017).

Assume a pass situation as presented in Figure 4.9. The main quantitative information of that pass model is the number of outplayed players. In addition,

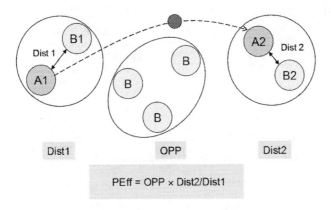

Figure 4.10 Pass efficiency *PEff* depending on the number of outplayed players *OPP* and the distances to closest opposing players at the moments of passing (Dist1) and receiving (Dist2).

qualitative information, that is also of tactical interest, is given by the respective distance between the involved players A1 and A2 and their closest opposing players B1 and B2 (see Figure 4.10).

The modelling idea contains three components:

> The number of outplayed opposing players is a quantitative component. A large number of outplayed players improves the efficiency of the pass.
>
> The distance between the passing player and their closest opponent is a qualitative component; shorter distances increase pressure, improve the technical evaluation and, therefore, the efficiency of the pass.
>
> The distance between the receiving player and their closest opponent is a qualitative component with tactical meaning; the greater that distance the more chance the receiving player has of getting the ball under control. Moreover, they will have more space for actions like dribbling, passing or shooting, i.e., larger distances to the closest opposing player improve the tactical options of the receiving player and, therefore, the efficiency of the pass.

Summing up, a mathematical formula of pass efficiency, PEff, has to model that efficiency:

> increases with the number of outplayed opposing players (OPP),
>
> increases with the distance (Dist2) between the receiving player and their closest opponent, and
>
> decreases with the distance (Dist1) between the passing player and their closest opponent.

The simplest formula, then, is given by PEff = OPP × Dist2/Dist1 (see Figure 4.10).

Creativity

A creative action can be defined as an action that, in the present situation, is rare and unexpected but successful (Memmert, 2011, 2015). In tennis, a ball played behind the back or through the legs would be an example. In soccer, technical feints are helpful to outplay an opposing defender. An old but good example for a tactical feint is the banana shot; at first, it looks like a pass to another player but, due to the spin on the ball, curves through the air and swerves towards the goal. Creative tactical actions are helpful in creating uncertainty on tactical plans and, therefore, are subject to training like technical feints (Memmert & Perl, 2009a). The problem is that systematic training of particular actions can develop them into behavioral patterns that become predictable and therefore not surprising. Thus, the idea is not to just train specific situations but to train the recognition of situations in which non-standard actions are required and how to choose an appropriate one. The complex demands on players in training based on this idea could be simulated by SOMs:

As has been pointed out, a network N can learn patterns of typical situations like formations of groups of players. In a similar way, a network N* can learn what the sets of most promising actions in a given situations are. Let a1 be an action out of a set A1 that is optimal in situation S1. Replacing action a1 by an action a2 does not necessarily reduce the expectation of success, as long as a2 belongs to A1. If, however, a2 belongs to a set A2 corresponding to a significantly different situation S2, then a2 could be rare and unexpected regarding situation S1, but it is questionable whether a2 is also successful in S1, i.e., was it creative or a fault? An appropriate algorithm working on N and N* can find answers to those questions and offer pairs of situations and actions that, although not coinciding optimally, would work successfully in the context of creativity.

This approach is much easier to describe than to get to work and is part of a research project being conducted by the authors of this chapter (Memmert, 2015; Memmert & Perl, 2009a; Memmert & Perl, 2009b).

Methodological experience

As noted above, the gap between idea and practice can be wide. And in particular, in modelling tactical behavior in the context of unknown intentions and incomplete data using mathematical methods and artificial neural networks, the gap between theory and practice can become significant. This gap between theory and practice is also apparent in the dialog between the researcher and the coach (Brefeld, Knauf & Memmert, 2016); not only are their languages different, but also the kind and complexity of information in which they are interested. Therefore, a transfer from research to practice is useful where complicated findings are translated to useful pieces of information. However, if simplified too much, a model does not represent the real system anymore, and the same with the information the model produces. This problem is discussed in more detail in the chapter dealing with KPIs. Therefore, in the following, the introduced methods and approaches of modelling are discussed in the context of representing reality.

Formations

One half in a soccer match has 2700 seconds plus, if necessary, additional time. This means that the number of observable formations is at least 2700 per half, if the time grid is defined by seconds. This, of course, is too much for informative tactical analysis. Therefore, modelling has to reduce this number by replacing the huge number of original formations by a small number of representative types, as has been done by means of artificial neural networks. The question is what "representative" means. There are at least two different meanings. On the one hand, it could mean "syntactic similarity" of formations belonging to the same type of geometric position relations. On the other hand it could mean "semantic similarity," where the position of one player in relation to the positions of the other members of the group is of tactical importance. For instance, in the context of tactical behavior, it could be important how far player 10 is in front of the attacking group. In this case, such detailed sub-formations could be of great interest, while typing by means of geometric patterns could be less important. This means that the kind and number of types depend on the goal of the analysis. Typing formations for the analysis of coincidence pairs and their success could be done with about 20 different types, while analysis of tactical behavior might need 40 or more types to avoid misinterpretation.

Moreover, if observing a play in a time grid of 1 second, there are about 2700 different formations. Compared to the 20 different formation types, this means that the mean expectation of an exact fit of formation type and original formation is about $20/2700 = 1/135$, i.e., less than 1%. Therefore, if using formation types as additional or even replacing information in an animation of the play, those 20 formation types could give the impression that formation modelling is worthless.

Based on our experience, formation modelling works well if it is restricted to tactical patterns and their sequences.

Summary: Modelling means abstraction, and abstraction simplifies handling but loses information. So a main challenge of modelling is to find the optimal balance between easy handling and useful information.

Voronoi cells

The same problem of easy handling vs. saving useful information appears in the case of modelling space control. As discussed on p. 83, the Voronoi cells of players in a real match depend on a lot of parameters and change continuously. If animation is the goal, modelling precision is a challenge and needs a lot of effort. If, however, the focus is on tactical behavior and processes, strong abstraction or idealization not only is permissible but necessary. Due to the fact that the precision of data recording is limited, the first step is to reduce spatio-temporal precision to grids of 1 meter and 1 second. The second step is to orientate the modelling to tactical aspects like changes and interactions of players and teams' space-control areas. Regarding players' control interaction during a time-depending process, 1 square meter more or less is not important. Corresponding to game theory, optimal and comparable moving parameters are assumed to provide answers also in worst-case situations. This sounds contradictory but can be understood as follows: If player B of team B has to move to optimize his space control against player A from team A, then the optimal move of player A, to which B must react, represents the worst case for B. Therefore, instead of weighing up player A's orientation, acceleration and speed at this moment, it makes more sense to simply assume they just move optimally. Moreover, regarding the team as a whole, small deviations of the players' control areas are compensated for by adding them to overall team control rates. Only this way is it possible to evaluate the success of the team's behavior in terms of measurable facts.

Summary: Abstraction can reduce information. But, depending on the modelling goal, idealization can improve the recording of information. So another main challenge of modelling is to find the optimal balance between useful and available information.

Pass analysis

In the context of modelling, pass analysis is one of the most interesting components of play analysis. The reason is that passes on the one hand build the dynamic base of the playing processes. On the other hand, from the viewpoint of modelling, passes are positioned between simple counting models and complex dynamic models, respectively depending on the context information taken into account, i.e., passes can be prepared as sub-systems from the complete system of processes. The quality of passes then can be measured depending on the information of the spatio-temporal context. Using appropriate analysis software, those analyses should not be too difficult.

It is much harder to measure the effect of such modelling approaches in order to assess a team's success based on those analysis results. The main methodological problem is that a direct coincidence between the quality of actions and the result of a play normally cannot be diagnosed. However, statistically, the quality of actions and processes is higher in the play of teams that perform well over time. This can be measured using correlation analysis. Unfortunately, the inverse direction does not hold, i.e., good actions do not guarantee good results, even if the correlation between the qualities of actions and results is high. The reason is that correlation analysis measures the quantity of similarity between the correspondent values of value sequences; it does not give logic information like "A implies B" if A correlates with B.

Nevertheless, even if "implies" does not work, correlation analyses give important information whether a model could be useful or does not work at all. If the resulting correlation coefficient is too small, the probability of not working is very high. If the coefficient is sufficiently great, this is not proof but at least an indicator of probably working. This is not a satisfying situation, but it is the only available approach in the case of modelling and assessing complex dynamic systems like plays. As a conclusion, one example of statistical model evaluation is presented below.

In a recent research project (Rein et al., 2017), vertical passes (i.e., passes from defensive players forward to offensive players) to the offense in an effort to create space control were analyzed. Passing behavior is a key property of successful performance in team sports. However, previous approaches have mainly focused on total passing frequencies in relation to game outcomes, providing little information about what actually constitutes successful passing behavior. This affects the application of these findings by practitioners. Here, by evaluating changes in majority situations in front of the goal and changes in space dominance due to passing behavior, two novel approaches to assessing passing effectiveness in elite soccer are represented. The majority situations are evaluated by calculating the number of opposing players between the ball carrier and the goal, while space dominance is assessed using Voronoi cells. The application of both methods to exemplary positional data from 12 German Bundesliga games demonstrates that they capture different features of passing behavior. The results further show that, on average, passes made from the midfield into the attacking area yield more effective results than passes within the attacking area. The present approaches provide interesting new avenues for more in-depth applications in future with immediate value for practitioners, for example in terms of tracking tactical schemes or performance, as the routines allow fast identification of individual passes. The routines are also applicable to other sports.

Results

Complexity of processes and tactical patterns

In order to model complex dynamic systems like soccer, a wide range of concepts and methods is necessary, stretching from simple counting of events and situations over short processes like passes to complex tactical dynamics like formations as parts of or contexts for processes.

Experienced coaches can read the play, i.e., recognize key events and key processes without calculating, simply imbedding their impressions in the background of the patterns they have learned. This way of compressing complex impressions into simple patterns is the most demanding challenge of modelling. The problem of pattern recognition is perfectly solved by brains—unfortunately without direct transfer to computers. Current research in artificial intelligence is trying to bridge that gap by developing, e.g., an artificial neural network as demonstrated in the case of formation pattern analysis.

Quantitative vs. qualitative information

From the researcher's point of view, tasks like generating and analyzing complex process patterns are much more exciting than counting numbers. However, science and life are different, and whether a model is useful or not is decided in the end by the aim of modelling that, in practice, is defined by the coach's demands. This means that a question like "quantitative or qualitative information" in applications in practice has to be decided in the context of what is needed to get helpful information and to reach decisions.

To sum up, modelling and analyzing sports like soccer can produce a lot of information that can be helpful for coaches, players, spectators, moderators and journalists—or even for bettors. But, unfortunately (or fortunately), real systems do not always behave like predicting models (Rein, Perl & Memmert, 2017).

In the last European Championship, for instance, a lot of matches ended as expected. But some of them did not, e.g., England, Belgium and Austria lost matches in which they were thought to be favorites, while Wales and Iceland as underdogs in European soccer became unexpected winners.

Without making models worthless: This is what makes tournaments and competitions so thrilling.

Example

The following example presents position-based modelling on three levels of increasing degree of difficulty (see Figure 4.11).

On the first level, the profiles of space control are calculated using the Voronoi approach (see p. 88). The black profile shows the time-depending control rates

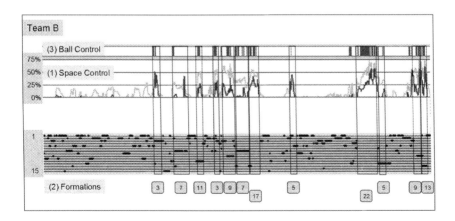

Figure 4.11 Three levels of modelling: (1) Quantitative space control, (2) patterns of tactical formations and (3) qualitative information about ball control (black: 16 m area; gray: 30 m area). Note that the formations "17" and "22" of team B are not contained in the set of the 15 most frequent ones.

in the 16 m area; the gray profile shows the respective rates in the 30 m area. The model is straightforward and relatively simple; only position data of players and a relatively simple algorithm are used.

On the second level, the time-depending offensive formations are added (see p. 88). The formation model uses a network approach and is significantly harder than the Voronoi approach. As the space-control model, the formation model only needs players' position data.

On the third level, the time-depending ball-control events are added (see p. 85). At first glance, this model seems to be the easiest one because it simply counts events. However, generating the regarding information from the position data is much harder because it not only needs additional information about the ball positions but also needs a sub-model to decide if and when the ball is controlled by a player.

Regarding modelling, the dynamics and tactics of the match at the three levels have quite different meanings. The third level of ball control models the technical basis of tactical processes. The first level of space control models the connection between technical abilities and tactical understanding, which is improved in connection with ball control. The second level of formations models the tactical background of the match. Together with information about ball control and space control, it facilitates evaluation of tactical abilities and behavior of a team (see Chapter 8 on KPIs).

Exercise

Ball recovery is one of the most important events in soccer because it presents a new chance to attack. The process starting with ball recovery gives important insight into the attacking efficiency of a team. Usually, the ball recovery process is described by a time sequence of four steps:

(1) ball recovery,
(2) temporal distance (e.g., 3 seconds) to the point of recovery,
(3) spatial distance of the ball (e.g., 30 meters) to the point of recovery, and
(4) final action (e.g., shot at goal) or ball loss.

In order to compare and assess such processes, those four steps have to be completed by context information on the respective situations like:

> the distances of players (which players, how close to what?),
> the speeds of players (which players, how close to what?),
> the formations (of which tactical groups of which team?), and
> space control (of which teams in which areas?).

Note: The more additional context information the higher the accuracy of description but, in turn, the information from statistical distributions decreases because the increasing number of items increases the number of similarity classes and so reduces the number of members per class. Therefore, concentrate on information that you think is most important. These are the tasks:

(1) Design a model of the recovery process by defining the steps and the context information of each step.
(2) Design an evaluation method based on your process model that helps to measure the process properties and/or quality.

References

Brefeld, U., Knauf, K. & Memmert, D. (2016). Spatio-temporal convolution kernels. *Machine Learning*, *102*(2), 247–273.
Carling, C., Wright, C., Nelson, L. & Bradley, P. (2014). Comment on 'Performance analysis in football: A critical review and implications for future research'. *Journal of Sports Sciences*, *32*, 2.
Duarte, R., Araújo, D., Freire, L., Folgado, H., Fernandes, O. & Davids, K. (2012). Intra- and inter-group coordination patterns reveal collective behaviors of football players near the scoring zone. *Human Movement Science*, *31*(6), 1639–1651.
Fonseca, S., Milho, J., Travassos, B. & Araújo, D. (2012). Spatial dynamics of team sports exposed by Voronoi diagrams. *Human Movement Science*, *31*(6), 1652–1659.
Frencken, W., Poel, H., Visscher, C. & Lemmink, K. (2012). Variability of inter-team distances associated with match events in elite-standard soccer. *Journal of Sports Sciences*, *30*(12), 1207–1213.
Gonçalves, B., Figueira, B., Maçãs, V. & Sampaio, J. (2014). Effect of player position on movement behaviour: Physical and physiological performances during an 11-a-side football game. *Journal of Sports Sciences*, *32*(2), 191–199.
Grunz, A., Perl, J. & Memmert, D. (2012). Tactical pattern recognition in soccer games by means of special self-organizing maps. *Human Movement Science*, *31*(2), 334–343.
Kim, S. (2004). Voronoi analysis of a soccer game. *Nonlinear Analysis: Modeling and Control*, *9*(3), 233–240.
Kohonen, T. (1995). *Self-organizing maps*. Berlin; Heidelberg; New York: Springer.
Lames, M. & McGarry, T. (2007). On the search for reliable performance indicators in game sports. *International Journal of Performance Analysis in Sport*, *7*(1), 62–79.
McGarry, T. (2009). Applied and theoretical perspectives of performance analysis in sport: Scientific issues and challenges. *International Journal of Performance Analysis in Sport*, *9*(1), 128–140.
Memmert, D. (2011). Sports and creativity. In M. Runco and S. Pritzker (Eds.), *Encyclopedia of creativity* (2nd edn.) (pp. 373–378). San Diego: Academic Press.
Memmert, D. & Perl, J. (2009a). Analysis and simulation of creativity learning by means of artificial neural networks. *Human Movement Science*, *28*, 263–282.
Memmert, D. & Perl, J. (2009b). Game creativity analysis by means of neural networks. *Journal of Sports Sciences*, *27*, 139–149.
Memmert, D. (2015). *Teaching tactical creativity in sport: Research and practice*. Abingdon: Routledge.
Memmert, D., Lemmink, K. & Sampaio, J. (2017). Current approaches to tactical performance analyses in soccer using position data. *Sports Medicine*, *47*, 1.
Perl, J. & Memmert, D. (2010). Game creativity analysis by means of artificial neural networks. *International Journal of Sport Psychology*, Supplement to *41*, 100–101.
Perl, J. & Memmert, D. (2011). Net-based game analysis by means of the software tool SOCCER. *International Journal of Computer Science in Sport*, *10*, 77–84.

Perl, J. (2015). Modeling and simulation. In A. Baca (Ed.), *Computer science in sport*. London; New York: Routledge.
Perl, J., Grunz, A. & Memmert, D. (2013). Tactics in soccer: An advanced approach. *International Journal of Computer Science in Sport, 12*, 33–44.
Rein, R. & Memmert, D. (2016). Big data and tactical analysis in elite soccer: Future challenges and opportunities for sports science. *SpringerPlus, 5*(1), 1410.
Rein, R., Perl, R. & Memmert, D. (2017). Maybe a tad early for a grand unified theory: Commentary on "Towards a Grand Unified Theory of sports performance" by Paul S. Glazier. *Human Movement Science, 56*(Pt A), 173–175.
Rein, R., Raabe, D. & Memmert, D. (2017). "Which pass is better?" Novel approaches to assess passing effectiveness in elite soccer. *Human Movement Science, 55*, 172–181.
Taki, T. & Hasegawa, J. (2000). Visualization of dominant region in team games and its application to teamwork analysis. In Proceedings, *Computer Graphics International, 2000* (pp. 227–235). IEEE.
Vilar, L., Araújo, D., Davids, K. & Bar-Yam, Y. (2013). Science of winning soccer: Emergent pattern-forming dynamics in association football. *Journal of Systems Science and Complexity, 26*(1), 73–84.

5 Modelling in the analysis of tactical behavior in team handball

Markus Tilp

Introduction

Besides technical skills and physical fitness, tactics is another major success factor in sports like team handball (Wagner, Finkenzeller, Würth & von Duvillard, 2014). Hence, coaches and scientists are seeking to better understand game tactics in order to improve game success. Classical approaches in the analysis of team handball tactical behavior by means of simply counting frequency of specific actions can reveal valuable information (Meletakos & Bayios, 2010; Meletakos, Vagenas & Bayios, 2011) but do not give full insight into the process of the game. Tactics in team handball and other sports is determined by the continuous interaction between the players of one's own and opposing teams. The understanding of these interactions, that consist of sequences of single actions, would help coaches and athletes to estimate their effect on the game or even to anticipate actions or events during the game. This would allow: a) estimating the success probability of different tactics, and b) determining and applying appropriate countermeasures. Therefore, several approaches to model the sequences of actions have been developed to map and analyze the process of the game.

One approach to model the sequence of actions in sports is the analysis of so-called temporal patterns (T-patterns; Magnusson, 2000; see Jonsson et al., 2010 for review). In this approach, single events (like passing, shooting, dribbling, etc.) in a sports game are arranged according to their temporal occurrence in the game in a first process step and subsequently scanned for similar sequential patterns. The main idea of the pattern detection is the temporal relationship between the events. Such T-patterns could be applied to sports like soccer, boxing, basketball or swimming.

An alternative approach to model action sequences in sports is to analyze the temporal behavior of exact position data (i.e., the position of the players/ball) and classify similar patterns by means of artificial neural networks (ANNs; Perl et al., 2013). In this approach, so-called self-organizing maps (SOMs) allow automatic clustering of complex data (e.g., position data of several players) to detect data patterns (e.g., tactical behavior). While pattern recognition by ANNs is widespread in many fields of science, in sports science, these techniques were first applied successfully in soccer (Memmert & Perl, 2009; Memmert et al., 2011).

Lately, our group has applied these methods in the analysis of action sequences in team handball. The aim of this chapter is to describe the challenges, the methods and examples of results of these applications in team handball.

Project/problem

The specific charm of team handball for spectators is the apparent unpredictability and the high variations during a game. Although various standard situations in the offensive tactics are part of this sport, these action sequences are never identical and always vary to some extent. This is a great challenge for game analysts who want to determine factors for success. One reason for the unpredictability of team handball is the possible interaction between 14 players participating in the game. This leads to almost unlimited different position and player combinations, which is challenging for coaches and sport scientists who want to determine the effectiveness of tactical behavior and to anticipate players' behavior. Counting the amount of exactly identical events will often lead to very low frequencies in the respective category. This makes the use of classical statistical methods difficult if not impossible because a minimum of frequencies in a category is often required to draw conclusions for sports practice. Furthermore, the number of actions, including passes, shots, etc., is enormous during a handball game. Meletakos and Bayios (2010) reported an average number of 58 goals/match in European National League men's handball in 2009. Taking into account that only about 60% of the attempts lead to a goal (Wagner, Gierlinger, Adzamija & von Duvillard, 2017) and that during each attempt several players are interacting, a lot of data is accumulated during a game. Taking into account these main challenges in the analysis of team handball tactics, methods are needed that can deal with: a) similarity (beyond equality), and b) huge amounts of data to analyze this complex sports game.

Concepts and methods of modelling

Different approaches have been used to simplify the huge amount of data in several types of sports. Scientists have applied, e.g., nonlinear statistical procedures like approximate entropy or relative phase analysis to overcome this challenge (Walter, Lames & McGarry, 2007; Sampaio & Maçãs, 2012; Goncalves, Figueira, Maçãs & Sampaio, 2014). Centroids and occupied surface area of teams or subgroups of teams were used to examine interaction between teams (Frencken, Lemmink, Delleman & Visscher, 2011; Duarte et al., 2012; Frencken, De Poel, Visscher & Lemmink, 2012). An interesting approach using Voronoi diagrams and nearest teammate distances was presented by Fonseca, Milho, Travassos and Araújo (2012) for futsal and later by Perl and Memmert (2016) for soccer. In these applications, Voronoi diagrams allow the spatial partitioning of the court into cells related to each player. Each cell includes the area of the court that is nearer to the related player compared to all other players. Summing up the amount of space occupied by all players of a team gives an estimate of the team's dominance on the court.

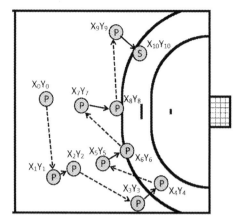

Figure 5.1 Schematic display of ball path in the last five passes prior to a shot at goal (P: pass; S: shot; solid line: running path; dashed line: passing path).
Source: adapted from Schrapf et al. (2017) with permission

In our working group, we have used ANNs (Perl, 2002) to model and analyze offensive and defensive behavior in team handball based on exact position data of specific game events. In order to record the sequence of game-relevant events, the positions of the shot and the preceding five passes were recorded by the custom-made MASA (Movement and Action Sequence Analysis) software (see Figure 5.4) that produces data vectors consisting of the coordinates of the events on the playing court. Each of these data vectors then represents a single offensive action sequence (see figures 5.1 and 5.2).

These data vectors (Figure 5.2, left panel) serve as input for an unsupervised ANN (Perl, 2002) that is able to detect frequent patterns, i.e., sets of similar action sequences, automatically. During the so-called training process of the network, similar input data vectors are related to the same or neighboring neurons on the output layer of the network. The different neurons and clusters of neighboring neurons in the output layer of the ANN represent different tactical offensive behavior (Figure 5.2, right panel). Once the ANN was "learned" and the input data was related to neurons and these were grouped to clusters, the semantic interpretation of the detected clusters was done by an expert. As an example, the expert could review all action sequences of a cluster on video and identify them as counter-attacks over the right wing of the court.

When an ANN has "learned" the relationship between input and output data after the training process, further data (vectors) can be presented to the net that are automatically related to the respective cluster. In our example, we could provide coordination data to the network that will relate it to specific offensive tactics.

Similarly, in order to detect defensive tactical patterns, the positions of the defensive team at the instant of the shot could be used as input data for the ANN

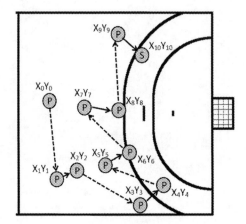

Figure 5.2 Data coordinates of player positions in X and Y direction (left panel) as input for an ANN. Each line represents an offensive action sequence comprising the shot and five preceding passes. Similar data vectors are automatically related to neighboring neurons on the output layer (right panel) of the artificial neural network-forming clusters (neurons in same color/with same number, right panel). Each cluster represents a tactical pattern that should be semantically interpreted by an expert. Hence, a cluster could represent, e.g., an attack from the left wing with a typical passing constellation.

and were scanned for common patterns (Schrapf, Alsaied & Tilp, 2017). Each cluster, automatically identified by the ANN, represents a different tactical defensive pattern of a team.

Methodological experience

ANNs have proven to be an adequate instrument to analyze action sequences in team handball based on position data. The automatic processing of position data is fast and more objective compared to manual video analysis. However, the analysis of tactical behavior in team handball with ANNs is related to several methodological challenges based on the complexity of the game and the way ANNs work.

One specific difficulty we observed was the recording of the position data. In general, this could be done by automated or manual video tracking or tracking of an active marker signal (Carling, Bloomfield, Nelsen & Reilly, 2008). In open-air sports like soccer, GPS tracking is an additional opportunity that, obviously, cannot be used in indoor handball. A restriction in position tracking in sports is that in most types of sports the use of active marker systems is prohibited during official games. Furthermore, an opposing team might not agree to wear your marker system during a game. Hence, until today, data recordings with active markers are restricted to training or friendly games in most types of sports although the signal stability and accuracy would be more user-friendly compared to video tracking. On the other hand, video tracking does not affect the athletes and is therefore applicable during official games and data from the opposing team can also be recorded. However, manual video tracking is labor-intensive and automatic tracking (e.g., Perš, Bon, Kovačič, Šibila & Dežman, 2002; Shortridge, Goldsberry & Adams, 2014) in difficult settings like a sports gym is still challenging for computer vision specialists. A specific difficulty is the camera perspective which leads to occlusions of players during the game. If players are occluded and the tracking software loses the defined player in a group of players, the tracking software has to be reinitiated. An alternative approach using cameras mounted on the ceiling would solve this problem for most situations (Perš et al., 2002). However, it is difficult to assess the technical quality of the players from the ceiling perspective. Therefore, we used eight cameras to cover the whole handball court from an elevated side-view perspective in our approach (see Figure 5.3). However, it was not possible to track all players during the game automatically and therefore we had to rely on position data at specific instants in time (e.g., when catching and passing the ball, ball throw, etc.). Based on the current technical developments, it can be expected that due to the miniaturization of markers that might be included in the fabric of sport clothes, position data will be easily available in the future. However, data availability of official games will still depend on the regulations of the respective type of sports. As soon as the position data is available, the calculations of the artificial network are done in seconds.

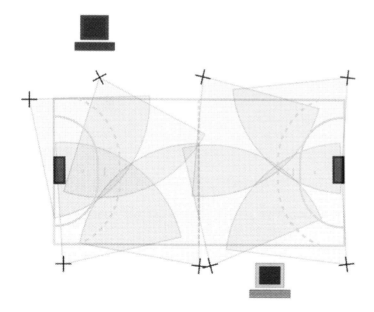

Figure 5.3 Computer positions, camera positions (indicated by "x") and camera perspectives during the video recording of handball games.

Another challenge is related to the application of ANNs which need an adequate amount of data in order to detect patterns. This means that you need to record a sufficient amount of sports data prior to the ANN analyses, which is sometimes difficult. To overcome this difficulty in our approaches we enlarged the data set by applying a noise of, e.g., 5% and multiplied the recorded position data. When the network receives enough input data, the network relates similar data to neighboring neurons and similar neurons are then combined into clusters. The necessary similarity thresholds (so-called "tolerance" for single neurons and "similarity resolution" for clusters) are given by the user who has to have expert knowledge to evaluate the discrimination process. However, once these thresholds are defined, the network discriminates the different action sequences automatically.

Following the pattern recognition it is often interesting how these patterns are related to success, i.e., if there are some patterns that are favorable to others. In team handball, a classical success criterion is a goal. Hence, if the offensive action leads to a goal it is positive and, if not, it is negative. This approach however neglects the individual performances of the attacker as well as the goalkeeper. This information cannot be included in the network process. A specific offensive action sequence could be perfect from a tactical perspective, i.e., puts the attacker in a very good position to place a shot, but the shot could be of minor quality (e.g., weak or directly at the goalkeeper) or the performance of the goalkeeper exceptional. In such a case, the action sequence would be assessed as not successful although the outcome was determined by individual performance. In order to overcome this shortcoming, we tried other definitions of success like distance to goal or distance to the nearest defender at the instant of the shot. It seems fair to assume that these parameters should be related to scoring rate. However, in our analysis, we could not determine such statistically significant relations. This was likely due to the fact that we included different teams with different players in the analysis. Hence, "individual player quality" might have affected results more than these shot conditions to reveal significant results. We recommend that future studies should investigate the relationship between the suggested parameters and classical goal-oriented models.

Results

ANNs have been used successfully to detect common playing patterns in team handball (Rudelstorfer et al., 2014). This allowed discriminating successful from unsuccessful offensive patterns in team handball (Schrapf & Tilp, 2013). The combination of offensive and defensive patterns enabled a deeper understanding of the interaction of teams (Schrapf et al., 2017). Lately, ANNs could even be used to objectively assess tactical training during competition (Hassan, Schrapf, Wael & Tilp, 2016) and to predict the shot position from precedent passing position data (Hassan, Schrapf & Tilp, 2017) in team handball. A detailed description of the results is given in the following examples:

Example 1: Determining offensive playing patterns in team handball by means of ANN

The aim of the study was to identify offensive patterns in team handball by means of ANNs (Schrapf & Tilp, 2013). Six games involving eight teams of the European Handball Federation's (EHF) Men-18 Championship were analyzed. Each game was captured by eight cameras (Figure 5.3) positioned to cover the whole playing area to calculate the players' positions (Rudelstorfer et al., 2014). All shots and passes were annotated using MASA software (see Figure 5.4).

The position data of the shot and five preceding passes was assembled to 612 action sequences and subsequently used as input for the ANN (DyCoN; Perl, 2002). In order to increase entropy for the training process of the ANN, original data was enlarged to 3060 data sets applying random noise of 15%. As a result, 42 different offensive strategies could be identified and related to clusters of the output layer of the network (Figure 5.2). Interestingly, 49% of all action sequences played were represented by only eight clusters, revealing common offensive patterns (Figure 5.5). The average deviation between the position of a single neuron and the corresponding original data was used as a benchmark for the quality of the assignment of action sequences to the clusters. In this approach, the deviation of only 1.2 m demonstrated a promising accordance of single action sequences with the related patterns. One important advantage of this approach compared to classical approaches is that context information is taken into account. Even when the final shot position of two action sequences is the same, the preceding actions might have been significantly different and lead to different defensive countermeasures.

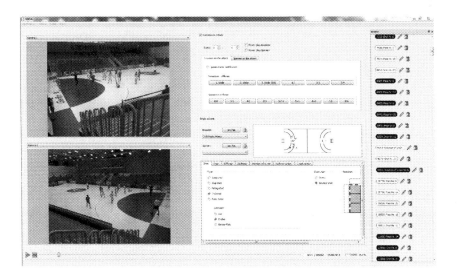

Figure 5.4 MASA user interface.

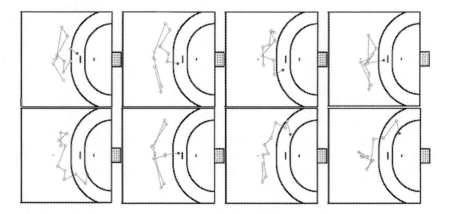

Figure 5.5 Illustration of the eight most common offensive patterns.
Source: adapted from Schrapf and Tilp (2013) with permission

Example 2: Analyzing tactical interaction of offensive and defensive teams in team handball by means of ANNs

Built upon our previous approaches on the offensive patterns, we aimed to model and analyze the interaction between offensive and defensive behavior of team handball teams in this project (Schrapf et al., 2017). Additional to the detection of offensive patterns, we identified defensive behavior by recording and analyzing the position of the defensive players at the instant of the shot by means of ANNs. Data of 12 games of an EHF Men-18 Championships were used to identify the playing patterns and to compare the success probabilities of different strategies by calculating odds ratios. The training process revealed 25 offensive and 13 defensive patterns out of the 7230 data sets, respectively. Subsequently, we analyzed the pattern combinations of these offensive and defensive strategies (Figure 5.6) that occurred eight times or more in the data. This was the case in 16 out of 325 (25x13) possible combinations. Additionally to the scoring rate, we assessed the success of an action sequence by the distance between shooting position and nearest defensive player as well as the distance to goal. Although no statistically significant relation between the cluster combinations and goal success rate was found, odds ratio analysis revealed that a specific defensive behavior was more effective than others against a specific type of offensive tactic. We observed, e.g., that the success of a specific offensive pattern (#2) was up 6.43 times higher against one defensive pattern compared to another frequently used defensive pattern (#1; see Schrapf et al., 2017 for details). Such information can be used to prepare a team with the appropriate defensive tactics against an opponent.

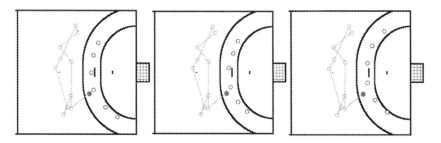

Figure 5.6 Combination of one specific offensive pattern (interconnected light gray circles) against different defensive patterns (dark gray circles).
Source: adapted from Schrapf et al. (2017) with permission

Example 3: Assessment of tactical training by means of ANNs

While there are plenty examples of game analysis of competition in different types of sport, there is a deficiency of game analysis research in the training process. In this project, we demonstrated that it is possible to evaluate tactical training in team handball by ANNs (Hassan et al., 2016). The aim of the project was to train offensive tactical patterns to a junior handball team and to evaluate the effect of the training process during a game. Twelve typical offensive patterns, i.e. running and passing paths of the shot and the previous five passes (Figure 5.1), were trained to the players. These patterns were previously identified by ANNs (Schrapf et al., 2017) from top-level U18 handball games and used as target patterns in the training process of the adolescent amateur handball team comprising 14 players (17+/-0.5 years, 7–10 years of training experience). Athletes trained the tactical patterns five times per week for 6 weeks. While in the first 2 weeks four target patterns were trained per training, this number was increased to six patterns starting from the third week. Tape was used to mark the target positions on the court in the training. The players were instructed to follow the running and passing paths as precisely as possible. Following the 6 weeks of tactical training, the team played an exhibition game which was video recorded. Position data of all offensive action sequences was digitized and stored. Subsequently, the data vectors of the action sequences were used as input for an ANN (NeuroDimension®, 2014). The network then tested the game patterns to check which patterns could be assigned to the previously trained 12 target patterns, revealing that 11 out of 12 target patterns could be detected during the game by the ANN. Twenty five (=58%) of the 43 recorded offensive action sequences could be assigned to a target pattern. This means that the players could apply the previously trained tactical patterns in the game to a great extent. Furthermore, we assessed the accuracy of the offensive patterns played during the game and the target patterns by calculating the mean difference between the related court positions. In mean, this distance was 0.49

(SD = 0.20) m. Although there are still some major drawbacks of the method, e.g., the time needed to undertake this type of analysis, it could be shown that ANNs are capable of assessing tactical training based on position data.

Example 4: Action predictions in handball by means of ANNs

Besides the modelling of interaction of players, the anticipation of behavior during sports is a fascinating challenge in game analysis as tactical anticipation is a key factor for success in sports. Human behavior is not deterministic and therefore impossible to anticipate 100%. However, one could predict actions with specific probabilities based on the procedural information of actions especially during standard situations in sports. Team handball is characterized by such standardized tactical patterns (Wagner et al., 2014) that can be identified by ANNs (Schrapf et al., 2013). Acknowledging the possibility to relate position data to a specific tactical pattern by ANNs, it seems feasible that even part of this information enables estimating its outcome with a certain probability. Therefore, the aim of this project was to estimate the position of the shot, based on position information of passing actions preceding the shot. Briefly, following the training of an ANN with 70% of the data and cross-validation with further 15% of the data, the remaining 15% was used for the analysis. Two approaches were undertaken to test our method on the data of 723 data sets, consisting of the position data of the shot and the five preceding passes (see Figure 5.1). In a first approach, the position information of all five passes and receives prior to the shot were used to predict the position of the shot. In a second approach, the receiving position of the shooting attacker and, subsequently, in another prediction step, the position of the shot itself, were estimated by the ANN. The ANN applied was a radial basis function network by NeuroDimension (2015). For technical details of the network, we refer to Hassan et al. (2017). To assess the accuracy of the estimations, the Euclidean distance between the calculated and the real shot position was calculated in both approaches. The difference between the predicted and real shot position was 1.20 (+/- 0.46 m) and 1.42 (+/-0.77 m) for the first and second approaches, respectively (see Figure 5.7).

Hence, with the information of the prediction analysis, the presumptive shot position of the attack could be anticipated with adequate accuracy. This allows for narrowing of the defense area to a meaningful size. We expect that teaching the defensive players the relationship between the initial actions and the most probable shot position in training could strengthen the defensive team.

In the described examples, we could show that ANNs can be used to identify offensive tactical patterns, to analyze the interaction between offensive and defensive teams, to evaluate tactical training during a game and to predict shot positions based on coordinates of preceding actions in team handball. Based on this experience, it is apparent that similar approaches can be applied in future projects in other sports like beach volleyball, volleyball, basketball or soccer. Position data of action sequences of the offensive and defensive teams could be integrated in an ANN analysis in basketball and soccer to study team interaction similar to the

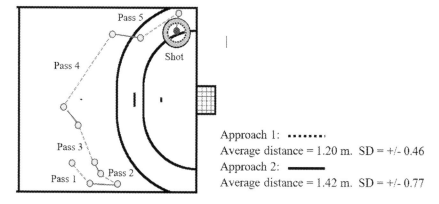

Figure 5.7 Mean areas around actual shot position based on mean distances between predicted and actual shot position of the two approaches (see text), respectively.
Source: adapted from Hassan et al. (2017) with permission

presented approaches in team handball. Preferred attacking tactics could be identified in beach volleyball based on the positions of the dig and the setter.

A first attempt to apply ANNs for predictions in volleyball has already been undertaken in a pilot study by our group (Schrapf, Hassan & Tilp, 2017b). By including the position of the setter at the instant of the serve, the position of the reception and the time elapsed between the reception and the setter's pass, we estimated the identity of the setter, the position of the attack and the type of the passing speed, i.e., the height of the pass. Following the training of the network with the described data, it was possible to correctly identify the setter with a mean accuracy of 55% (±14.7), to predict the position of attack with 57.6% (±16.1) and the passing speed with 83.3% (± 12.5), all of which were significantly higher than by random prediction.

Exercises

- Find other types of sports where position data can be used to model action sequences and be analyzed with ANNs.
- Which position data could give insights into the important action sequences during a sports game?
- Which events could be interesting to predict (by ANNs) in other types of sports?

References

Carling, C., Bloomfield, J., Nelson, L. & Reilly, T. (2008). The role of motion analysis in elite soccer: Contemporary performance measurements techniques and work rate data. *Sports Medicine* (Auckland, N.Z.), *38*(10), 839–862.

Duarte, R., Araújo, D., Freire, L., Folgado, H., Fernandes, O. & Davids, K. (2012). Intra- and inter-group coordination patterns reveal collective behaviors of football players near the scoring zone. *Human Movement Science*, *31*, 1639–1651.

Fonseca, S., Milho, J., Travassos, B. & Araújo, D. (2012). Spatial dynamics of team sports exposed by Voronoi diagrams. *Human Movement Science*, *31*, 1652–1659.

Frencken, W., Lemmink, K., Delleman, N. & Visscher, C. (2011). Oscillations of centroid position and surface area of soccer teams in small-sided games. *European Journal of Sport Science*, *11*(4), 215–223.

Frencken, W., De Poel, H., Visscher, C. & Lemmink, K. (2012). Variability of inter-team distances associated with match events in elite-standard soccer. *Journal of Sports Science*, *30*(12), 1207–1213.

Gómez, M., Lorenzo, A., Ibáñez, S., Ortega, E., Leite, N. & Sampaio, J. (2010). An analysis of defensive strategies used by home and away basketball teams. *Perceptual and Motor Skills*, *110*(1), 159–166.

Hassan, A., Schrapf, N., Wael, R. & Tilp, M. (2016). Evaluation of tactical training in team handball by means of neural networks. *Journal of Sports Sciences*, *35*(7), 642–647.

Hassan, A., Schrapf, N. & Tilp, M. (2017). The prediction of action positions in team handball by nonlinear hybrid neural networks. *International Journal of Performance Analysis in Sport*, *17*(3), 293–302.

Jonsson, G., Anguera, M., Sánchez-Algarra, P., Olivera, C., Campanico, J., Castañer, M., ... Magnusson, M. (2010). Application of T-Pattern detection and analysis in sports research. *The Open Sports Sciences Journal*, *3*, 95–104.

Magnusson, M. (2000). Discovering hidden time patterns in behavior: T patterns and their detection. *Behavior Research Methods, Instruments & Computers*, *32*, 93–110.

Memmert, D. & Perl, J. (2009) Analysis and simulation of creativity learning by means of artificial neural networks. *Human Movement Science*, *28*, 263–282.

Memmert, D., Bischof, J., Endler, S., Grunz, A., Schmid, M., Schmidt, A. & Perl, J. (2011). World-level analysis in top level football analysis and simulation of football specific group tactics by means of adaptive neural networks. In Chi-Leung Hui (Ed.), *Artificial Neural Networks – Application* (pp. 3–12). Rijeka: InTech.

Meletakos, P. & Bayios, I. (2010). General trends in European men's handball: A longitudinal study. *International Journal of Performance Analysis of Sport*, *10*(3), 221–228.

Meletakos, P., Vagenas, G. & Bayios, I. (2011). A multivariate assessment of offensive performance indicators in men's handball: Trends and differences in the World Championships. *International Journal of Performance Analysis of Sport*, *11*(2), 284–294.

NeuroDimension®. (2014). *NeuroSolutions*. Retrieved June 1, 2018 from www.neurosolutions.com/neurosolutions/.

Perl, J. (2002). Game analysis and control by means of continuously learning networks. *International Journal of Performance Analysis of Sport*, *2*, 21–35.

Perl, J. & Memmert, D. (2016). Soccer analyses by means of artificial neural networks, automatic pass recognition and Voronoi-cells: An approach of measuring tactical success. In P. Chung, A. Soltoggio, C. Dawson, Q. Meng & M. Pain (Eds.), Proceedings of The 10th International Symposium on Computer Science in Sports (ISCSS). *Advances in Intelligent Systems and Computing*, *392*, (1st ed., pp. 77–84). Cham: Springer.

Perl, J., Tilp, M., Baca, A. & Memmert, D. (2013). Neural networks for analysing sports games. In T. McGarry, P. O'Donoghue & J. Sampaio (Eds.), *Routledge handbook of sports performance analysis* (pp. 237–247). New York: Routledge.

Perš, J., Bon, M., Kovačič, S., Šibila, M. & Dežman, B. (2002). Observation and analysis of large-scale human motion. *Human Movement Science*, *21*(2), 295–311.

Rudelstorfer, P., Schrapf, N., Possegger, H., Mauthner, T., Bischof, H. & Tilp M. (2014). A novel method for the analysis of sequential actions in team handball. *International Journal of Computer Science in Sport*, *13*(1), 69–84.

Sampaio, J. & Maçãs, V. (2012). Measuring tactical behaviour in football. *International Journal of Sports Medicine*, *33*, 395–401.

Sampaio, J., Lago, C., Casais, L. & Leite, N. (2010). Effects of starting score-line, game location, and quality of opposition in basketball quarter score. *European Journal of Sport Science*, *10*(6), 391–396.

Shortridge, A., Goldsberry, K. & Adams, M. (2014). Creating space to shoot: Quantifying spatial relative field goal efficiency in basketball. *Journal of Quantitative Analysis in Sports*, *10*(3), 303–313.

Schrapf, N. & Tilp, M. (2013). Action sequence analysis in team handball. *Journal of Human Sport and Exercise*, *8*(3), 615–621.

Schrapf, N., Alsaied, S. & Tilp, M. (2017). Tactical interaction of offensive and defensive teams in team handball analysed by artificial neural networks. *Mathematical and Computer Modeling of Dynamical Systems: Methods, tools and applications in engineering and related sciences*, *23*(4), 363–371.

Schrapf, N., Hassan, A. & Tilp, M. (2017b). Analysis of setters' passing behavior within complex 1 in volleyball by means of artificial neural networks. In A. Ferrauti, P. Platen, E. Grimminger-Seidensticker, T. Jaitner, U. Bartmus, L. Becher, ... E. Tsolakidis, (Eds.), *ECSS Book of abstracts* (p. 462). Cologne: Sportools.

Wagner, H., Finkenzeller, T., Würth, S. & von Duvillard, S. (2014). Individual and team performance in team-handball: A review. *Journal of Sports Science Medicine*, *13*, 808–816.

Wagner, H., Gierlinger, M., Adzamija, N. & von Duvillard, S. (2017). Specific physical training in elite male team handball. *The Journal of Strength and Conditioning Research*, *31*(11), 3083–3093.

Walter, F., Lames, M. & McGarry, T. (2007) Analysis of sports performance as a dynamic system by means of relative phase. *International Journal of Computer Science in Sports*, *6*(2), 35–41.

6 Basketball

Jaime Sampaio, Bruno Gonçalves, Nuno Mateus, Zhang Shaoliang and Nuno Leite

Introduction

Performance in high-level basketball is a very complex process to understand, mainly due to its dependency on a substantial number of dynamical interactions between technical, tactical, fitness and anthropometric characteristics of players (Sampaio, Ibáñez & Lorenzo, 2013). In general, available research has used game-related statistics to identify the most important performance indicators (Ittenbach, Kloos & Etheridge, 1992; Trninic, Perica & Dizdar, 1999; Karipidis, Fotinakis, Taxildaris & Fatouros, 2001; Sampaio & Janeira, 2003) and, afterwards, has modeled performance with the help of several situational variables such as game location, quality of opponent or scoreline (Gomez, Lago-Peñas & Pollard, 2013a; Gomez, Lorenzo, Ibáñez & Sampaio, 2013b). In addition, it has been possible to understand the effects of using alternative defensive strategies, such as individual or zone defense (Gomez et al., 2010) or to comprehend how the defensive pressure (full- or half-court press) affects technical actions (Sampaio, Lago, Casais & Leite, 2010). Results available are mainly based on samples from European competitions and suggest that basketball performance depends offensively on shooting field goals and defensively on securing defensive rebounds (Karipidis et al., 2001; Sampaio & Janeira, 2003; Malarranha, Figueira, Leite & Sampaio, 2013). However, in more specific contexts such as closely contested games, fouls and free throws might exhibit higher importance (Kozar, Vaughn, Whitfield, Lord & Dye, 1994; Sampaio & Janeira, 2003). Most of these results can also be identified during the play-off stages of the competition where players are required to respond to critical moments and circumstances that are substantially different from the regular season (García, Ibáñez, Martinez De Santos, Leite & Sampaio, 2013).

The other game-related statistics such as offensive rebounds, turnovers, steals or assists, seem to be inconsistently reported as discriminators between winners and losers. There are also results that relate the best performances to assists, steals and blocks, denoting the importance of passing skills as well as the exterior and interior defensive intensity (Ibáñez et al., 2008). Most recently, players' external load started to be measured more accurately, providing important information not only about performance (Sampaio et al., 2015), but also about the risk of injury (Caparrós et al., 2017).

Playing time and player status have also drawn attention from researchers. Unsurprisingly, errors seem to be the most important factor when contrasting more important and less important players, with fewer errors being made by more important players (Sampaio, Lago, Casais & Leite, 2010). Recent studies performed with NBA (National Basketball Association of the United States of America) players added that playing time seems to have a key role in enhancing lower-body power, repetitive jump ability and reaction during the competitive season (Gonzalez et al., 2013).

One of the most recent trends in sports science is the individualization of all protocols for training and monitoring, especially in the NBA, where the highly congested schedule can directly affect players' performance. In fact, it seems clear that each player responds individually to the stress of practice and competition, thus there seems to be a need to use updated performance models to act as starting points in players' preparation, as well as fine-tuned measurements of their performance. Several approaches were employed in NBA games attempting to fine-tune the models that can serve this purpose for high-level basketball. For example, it has been shown that NBA all-star players had better field goal percentages within 12 feet of the basket and it was possible to identify groups of performers, particularly related to roles of scoring, passing, defensive and all-round tasks in the game (Sampaio et al., 2015). These performance differences are influenced by players' court position and have a direct repercussion on their playing time (Mateus et al., 2015). Nonetheless, caution is advised in assuming any direct relationship between players' court position and game statistics, particularly in guards and centers, since the continuous evolution of the game and the players' athleticism levels probably increase within-position variability (Mateus et al., 2015). The problem of individualization contrasts with the obvious need to consider the players as part of a team. On one hand, the coaches need to individualize preparation but, on the other, they need to pull the team together. Available research tries to minimize the individualization problem by grouping players using their game-specific positions, using three or five different levels (Mateus et al., 2015; Pojskic, Separovic, Uzicanin, Muratovic & Mackovic, 2015; Calleja-Gonzalez et al., 2016; Shaoliang et al., 2017). However, considering the above-mentioned evidence, there is always criticism of these standardized and subjective grouping procedures. Therefore, the possibility of using methodologies that are capable of extending the stage of playing-position analysis and providing information about individual performance might be very helpful in revealing new insights about the true value of players. Finally, evaluating players by individual performance profiles is overdependent on their scoring capabilities (Mertz et al., 2016), that is, point production might obscure all the other positive and negative actions of the players. This way, basketball analysis could provide clearer results by using different levels of point production in the games. Summing up, the current exercise of modelling will try to provide an individualized perspective of performance in basketball, bearing in mind differences in point production of the players and their specific positions.

Problem

Basketball is a team sport where box-score statistics are often used to help identify the reasons that explain the game's outcome. However, the box scores only contain information that describe the frequency of actions performed by players of both teams in a game. This description can often be considered as a complete representation of the game, however measuring performance is still a never-ending challenge for basketball coaches, trainers or recruiters. Player-tracking technology is one of the most recent technological advances in basketball. Powerful computer vision systems have been designed with fine-tuned algorithms capable of tracking with relatively high accuracy the players' positioning and, subsequently, all derived variables such as distance covered and speed. A very interesting advance is the combination of performance dimensions, such as physiological and technical, throughout the usual notational analysis. Therefore it is possible, for example, to analyze the distance covered by the players when the team is attacking and when the team is defending. Studies focused on positional-derived variables in basketball are still limited to small samples of young basketballers, by examining physical demands (Ben Abdelkrim, El Fazaa & El Ati, 2007), effects of defensive pressure on movement behavior (Leite et al., 2014) and effects of activity workload in tactical performances (Sampaio, Gonçalves, Rentero, Abrantes & Leite, 2014).

Nowadays, there is an urgent need to develop and implement the use of adequate performance indicators. In fact, sports organizations collect substantial amounts of data; often, however, there is no certainty that these measures are really linked with performance and with success. The gathered data needs to be transformed into performance indicators, with the aim of producing measures linked to performance processes and outcomes. Basketball performance is complex in the way that interactions between players and opponents provide for emergent behavior to occur. Conversely, performance is also dynamic, meaning that most interactions can be time dependent and, finally, performances have substantial nonlinear properties, because the output of the system is rarely directly proportional to its input. Therefore, theoretically, the performance indicators should be able to capture most global or partial aspects of these complex, dynamic and nonlinear properties as required.

Current research has used several criteria to approach validity of performance indicators by using elementary and advanced statistical techniques to compare winners against losers (Gomez, Lorenzo, Sampaio, Ibáñez & Ortega, 2008). The actual holistic understanding of the game often requires analyzing several sets of (dependent) variables simultaneously and, therefore, the use of statistical procedures such as multivariate analysis of variance, principal components analysis, discriminant analysis, cluster analysis or artificial neural networks may be the most appropriate (Bracewell, 2003).

In all professional and developmental basketball leagues, the data-gathering process is standardized and regulated by the operational definitions and criteria published in the Basketball Statisticians Manual (FIBA, 2016). This is a very important point that ensures intra- and inter-operator reliability; however, by having reliable data, the assumption of validity is not necessarily comprised, i.e., although the data may be consistent it is not obvious that most of these actions really transmit information about performance. Nevertheless, the game actions presented in the box scores and defined in this official manual are the following: free throws, field goals, rebounds, assists, steals, turnovers, blocked shots and fouls. Using these actions, there has been an intensive search for an accurate group of performance indicators (Oliver, 2004), so it has been suggested that the best breakdown of offensive and defensive performances can be obtained by analyzing four factors in the following order of importance: (1) effective field-goal percentage, (2) offensive rebounding percentage, (3) turnovers per ball possession and (4) free-throw rate (Oliver, 2004; Kubatko, Oliver, Pelton & Rosenbaum, 2007). Offensively, a team wants to minimize turnovers per possession and maximize all the other factors. These factors are not all equivalent, meaning that, for NBA games, the relative weights of these are approximately 10, 6, 3 and 3, respectively for each factor, but that might change in other competitions (Kubatko et al., 2007).

In a basketball team, point scoring often obscures other key determinants of collective success that, supposedly, can be measured using other box-score statistics. An alternative solution to refining performance metrics can be proposed by classifying players into different levels of point production. This way, lower-scoring players would improve their chances of having their performance evaluated and recognized and, later, coaching staff can access objective data to provide suggestions for improvement. Following up this idea of analyzing different levels of scoring production, there is a clear need to identify the subset of other game statistics that discriminate these levels of point production, when accounting for specific game positions (guards, forwards and centers) that probably have different action requirements. At the end of this process, an optimized model can be built by comparing players' performances in a multidimensional space containing the most important variables and accounting for their point production and specific court position. Obtained results would likely help us understand the complex interactions that determine successful performances in the game.

Concepts and methods of modelling

A possible approach to the problem will require the sequential usage of three different techniques: (1) cluster analysis, (2) discriminant analysis and (3) nearest neighbor analysis.

Figure 6.1 presents the depiction of a possible approach. The first step to accomplish the analysis consisted of using a two-step cluster analysis, a technique designed to reveal natural groupings within large data sets that would otherwise not be apparent. Log-likelihood is used as a distance measure and Schwarz's Bayesian Criterion as a clustering criterion. The log-likelihood places a probability distribution on the variables in a way that continuous variables are assumed

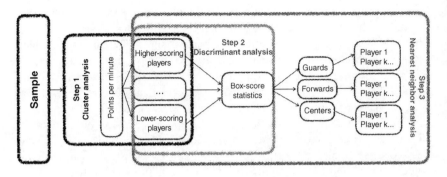

Figure 6.1 Depiction of the three steps of a possible approach.

to be normally distributed, while categorical variables are assumed to be multinomial. In addition, all variables are assumed to be independent; however, empirical internal testing indicates that the procedure is fairly robust to violations of these assumptions. This method is a scalable cluster analysis algorithm designed to handle very large data sets (both continuous and categorical variables or attributes). The two steps consist of pre clustering the cases into many small subclusters and then clustering the sub-clusters resulting from the pre-cluster step into the desired number of clusters (Zhang, Ramakrishnon & Livny, 1996; Chiu, Fang, Chen, Wang & Jeris, 2001).

The descriptive discriminant analysis aims to identify a smaller subset of variables able to discriminate between *a priori* defined groups, in this case, using the categories obtained by the previous cluster analysis as the levels from the grouping variable. This technique seeks out a linear combination of large amounts of variables for each group that maximizes their differences for proper classification. The analysis provides a classification function that determines which groups an individual belongs to. All independent variables enter the model together and the prior probabilities of classification were entered according to the size of the groups (Cooley & Lohnes, 1971; Tatsuoka, 1971). The obtained classification function coefficients can be used to develop predictive equations from the level of scoring in each playing position. The most important variables are identified by the structure coefficients that reached values above $|0.30|$ and the precision of the models are evaluated by the leave-one-out method (Pedhazur, 1982).

Nearest neighbor analysis (kNN) is a non-parametric technique that can be used for classification where the input consists of the k-closest training examples in the feature space. A case is classified by a majority vote of its neighbors, with the case being assigned to the class most common among its k-nearest neighbors (k is a positive integer, typically small). For example, if k = 1, the case is simply assigned to the class of that single nearest neighbor. The kNN can use as features the most important variables identified by the discriminant analysis and as targets the levels identified by the cluster analysis. The number of neighbors (k) can be automatically determined, after testing a minimum of 3 and a maximum of 20. Usually, only the model with less error rate is selected and presented. The Euclidean metric is used to compute distances weighted by the importance of the features (Arya and Mount, 1993; Cunningham & Delaney, 2007; Friedman, Bentley & Finkel, 1977). In addition, there is a training and a holdout partition that comprised 70% and 30% of the sample, respectively.

Methodological experience

Data analysis was performed in the NBA database containing all players' performances from the 2015–2016 regular season. The raw data file originally contained 27,966 cases but was transformed in order to aggregate each player performance into a single profile row by averaging the values from all the available game variables. This can be done with these average values, but researchers can also use other approaches such as standard deviations, coefficient of variation, 25th percentile or 75th percentile. For example, using the coefficient of variation can be appropriate to examine variability and obtain information about how players adapt their performance to different environments. Additionally, using the 25th percentile can allow to obtain information about the worst performance scenarios, whereas using the 75th percentile can allow to describe the best performance scenarios.

From the original database, there were 43 players averaging less than one-eighth of play (6 min) per game over the season and they were removed from the database. The final number of players in the file was 488 from the 30 participating teams.

Figure 6.1 presents the depiction of the proposed approach. The two-step cluster analysis was carried out using the continuous variable points scored per minute of play and the number of clusters was determined automatically. This procedure enables to classify players according to their different levels of scoring and, afterwards, this new categorical variable allows to understand better players' performance contributors.

The discriminant analysis was the second step in the analysis, allowing to identify the smaller subset of game variables that can discriminate players' different levels of scoring, but also according to their court positions (guards, forwards and centers). The independent variables comprised a set of 13 game variables:

- 2-point field-goal percentages,
- 3-point field-goal percentages,
- free-throw percentages,
- average defensive rebounds per game,
- average offensive rebounds per game,
- fouls committed,
- turnovers,
- blocks,
- assists,

114 Sampaio et al.

- steals,
- speed,
- distance covered, and
- touches per minute of play.

At the end of this stage, it was possible to identify the most powerful game actions in discriminating different levels of scoring for guards, forwards and centers. Also, the obtained equations can help in classifying new cases into the scoring groups.

The third step, nearest neighbor analysis (kNN), consisted of classifying the players' performances based on their similarity to other players and was performed for each playing position using the most important variables identified by the discriminant analysis and the levels of scoring determined by the cluster analysis. The kNN technique allows to identify the most similar players using the most important game variables according to the level of scoring and playing positions. All the analyses were performed using IBM SPSS software (release 22.0, SPSS Inc., Chicago, IL).

Results

Step 1: Cluster analysis—determining different levels of scoring

In the first step, the obtained results provided an automatically determined solution with two clusters with a good silhouette measure of cohesion and separation (average silhouette = 0.7). A new categorical variable was saved in the database with these classification results. For designation purposes, one of the clusters was labeled as "lower-scoring players" and gathered 258 players (53%) averaging 0.30 ± 0.06 points per minute (range 0–0.38) and the other cluster was labeled as "higher-scoring players," gathering 230 players (47%) that averaged 0.49 ± 0.09 points per minute (range 0.39–0.85). Cross-tabulations between the obtained cluster solution and playing positions produced very similar distributions (lower-scoring guards 52.9%; higher-scoring guards 47.1%; lower-scoring forwards 54.9%; higher-scoring forwards 45.1%; lower-scoring centers 51.3%; higher-scoring centers 48.7%).

Step 2: Discriminant analysis—identifying a smaller subset of important variables

In the second step, the obtained discriminant functions were all significant ($p \leq 0.001$) with chi-square values of 119.7 for guards, 105.9 for forwards and 42.8 for centers. The canonical correlations were 0.66 for guards, 0.66 for forwards and 0.68 for centers. Table 6.1 presents the linear classification function coefficients and structure coefficients for each playing position and scoring level. Concerning the structure coefficients, there were clear differences between

Table 6.1 Linear classification function coefficients for higher- and lower-scoring players and structure coefficients for each position. The case example provided in the second column refers to the game average performance of the guard Stephen Curry.

Variables	Guards				Forwards				Centers		
	Case Example	Lower–scoring	Higher–scoring	SC	Lower–scoring	Higher–scoring	SC		Lower–scoring	Higher–scoring	SC
Assists	6.39	0.48	−0.01	0.36	1.08	−0.05	0.35		8.75	7.07	0.32
Blocks	0.21	−7.01	−9.33	0.03	4.21	4.54	0.37		−19.72	−20.54	0.16
Dist covered	0.07	1920	1867	−0.20	2818	2837	−0.22		1155	928	−0.28
Def rebounds	4.61	1.57	1.23	0.32	1.57	1.25	0.55		12.41	12.30	0.33
2-pt field goals	48.78	0.32	0.43	0.55	1.01	1.09	0.44		0.89	0.99	0.17
Free throws	81.42	0.38	0.46	0.74	0.02	0.11	0.73		−0.37	−0.27	0.63
Off rebounds	0.85	−4.87	−5.16	0.13	−13.11	−12.59	0.39		−26.08	−26.91	0.16
3-pt field goals	43.41	0.06	0.09	0.37	−0.06	−0.01	0.18		−0.75	−0.71	0.31
Com fouls	2.06	4.23	3.15	0.20	7.10	6.31	0.33		7.04	6.68	0.25
Speed	4.28	68.78	68.78	−0.18	72.53	70.90	−0.22		163.56	163.14	−0.25
Steals	2.01	−3.50	−3.80	0.28	−17.85	−18.98	0.23		−24.24	−23.10	0.34
Touches	2.48	−1.29	−0.74	0.21	19.61	22.79	0.40		−14.68	−10.33	0.37
Turnovers	3.47	−4.66	−2.89	0.52	7.45	8.96	0.52		26.31	28.00	0.52
(Constant)		−228.19	−231.31	—	−286.84	−292.28	—		−386.53	−382.93	—
Discriminant score		696	708								

playing positions. Nevertheless, there were variables achieving high discriminant status for all positions, such as the free throws and turnovers, and more moderate variables, such as the assists and the defensive rebounds. Interestingly, distance covered and speed were the only variables with no substantial discriminant status. In addition, the guards' discriminant variables were related to field goals (2- and 3-point), the forwards to many other variables (e.g., free throws, defensive rebounds, ...) and the centers to 3-point field goals, touches and steals (see Table 6.1).

The linear classification function coefficients allow to calculate the discriminant scores for a given case (see Table 6.1). The case example data (Stephen Curry) was used to calculate the discriminant score for each possibility of classification (Discriminant score = Variable1*Coefficient1+ Variable2*Coefficient2 ... + constant). Afterwards, the higher discriminant score will correspond to the higher probability of being classified in that group. In the case example, the player will be classified as a higher-scoring guard (Discriminant score$_{lower-scoring}$ = 696 vs Discriminant score$_{higher-scoring}$ = 708).

The leave-one-out classification allowed to test the accuracy of these models and, for each playing position, the obtained results were high (guards = 80.2%, forwards = 80.1% and centers = 79.5%).

The discriminant scores can be used to analyze several aspects of performance. For example, Figure 6.2 presents the distribution of discriminant scores per player and per team, according to high- and low-scoring players. Each of the panels allows to identify the teams' homogeneity in each situation. In high-scoring players, for example, SAC has one player that is clearly detached positively from the rest of the team, whereas LAC has one player that detaches negatively (Figure 6.2). WAS seems a very homogeneous team and OKC players are much more spread out, showing more inter-player differences in performance. In low-scoring players, several teams show at least one player detached from the rest (BOS, CHI, HOU, MEM, MIN, NOP, ORL, PHI, SAC, TOR). The player from NOP had the smallest discriminant score in the whole sample.

Complementary to the previous information, Figure 6.3 depicts the teams' constitution profile, according to high- and low-scoring players for each playing position. This visualization allows to understand how the previous discriminant scores for each specific position help to identify teams' relative strengthens and weaknesses. For example, SAC, OKC, MEM and GSW are identified as teams most represented by discriminate scores from high-scoring centers. Conversely, there are no players from CLE, DET, LAL, NYK and POR in this group, meaning that centers from these teams have different playing profiles. By analyzing the forwards' performance, NOP, NYC and OKC were the top teams from the high-scoring group and TOR, SAC, ORL the top teams from the lower-scoring group. This last result denotes these teams' problems in scoring and performing from these positions, when compared to the other teams. Finally, GSW, HOU and PHI were the teams with higher performance in guards from the high-scoring group and NOP denoting performance problems as seen by the lower-scoring

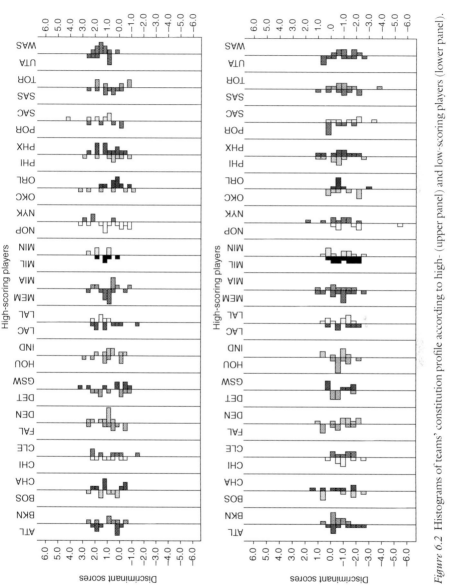

Figure 6.2 Histograms of teams' constitution profile according to high- (upper panel) and low-scoring players (lower panel).

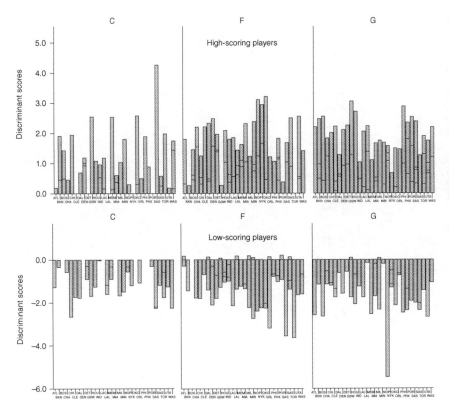

Figure 6.3 Teams' constitution profile according to high- (upper panel) and low-scoring players (lower panel) for each playing position: centers (C), forwards (F) and guards (G). The discriminant scores are the sum for the players in the specific positions.

discriminant scores. This analysis can be useful in evaluating teams by their complementary performance profiles, as well as in scouting opponents and recruiting new prospects.

Step 3: Nearest neighbor analysis—identifying the most similar players

In the third step, the analysis from the guards yielded good percentages of correct classifications for training (76.4%) and holdout (70.0%). Nevertheless, lower-scoring players had better percentages of correct classifications (83.3% vs. 68.5% in training and 78.6% vs. 62.5% for holdout). Figure 6.4 presents a lower-dimensional projection of the predictor space that contains six predictors ($k = 7$, error = 0.22). In addition, the quadrant map presented in Figure 6.5 compares

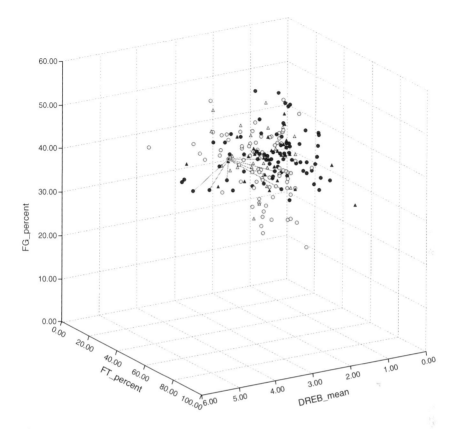

Figure 6.4 Lower-dimensional projection of the predictor space containing six predictors: free throws (FT_percent), 2-point field goals (FG_percent), turnovers, 3-point field goals, assists and defensive rebounds (DREB_mean). The focal case is the guard with higher points scored per minute of play (Stephen Curry). Circles represent training cases, and triangles holdout cases. The higher-scoring players have dark filling and the lower-scoring players light filling.

the most point-productive guard (focal case: Stephen Curry) with their seven neighbors when performing in the most important predictors.

The analysis from the forwards yielded good percentages of correct classifications for training (75.8%) and holdout (76.2%). Again, the lower-scoring players had better percentages of correct classifications (90.4% vs. 56.4% in training and 93.5% vs. 59.4% for holdout). Figure 6.6 presents a lower-dimensional projection of the predictor space that contains nine predictors (k = 4, error = 0.20). In addition, the quadrant map from Figure 6.7 compares the most point-productive forward (Kevin Durant) with their four neighbors.

The analysis from the centers yielded good percentages of correct classifications for training (69.1%) and holdout (82.6%). In contrast to the other

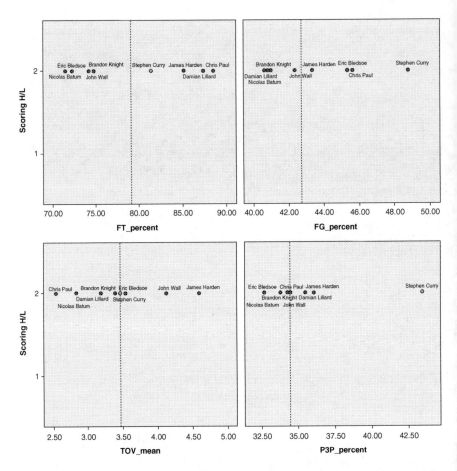

Figure 6.5 Target values by predictors for initial focal cases and nearest neighbors. The focal case is the guard with higher points scored per minute of play (Stephen Curry). Circles represent training cases and triangles holdout cases. The higher-scoring players were allocated with a score of 2 and lower-scoring players with a score of 1 (dummy variable).

playing positions, the higher-scoring players had better percentages of correct classifications (67.9% vs. 70.4% in training and 75.0% vs. 90.9% for holdout). Figure 6.8 presents a lower-dimensional projection of the predictor space that contains seven predictors (k = 11, error = 0.29). The quadrant map presented in Figure 6.9 refers to the comparison from the center with higher values of points scored per minute of play (DeMarcus Cousins) and their k-neighbors concerning four predictors. There was one case of a player classified as low-scoring (Rudy Gobert) that reached similar performances of the other peers.

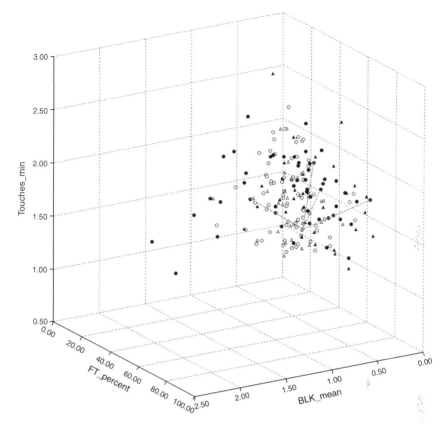

Figure 6.6 Lower-dimensional projection of the predictor space containing nine predictors: free throws (FT_percent), defensive rebounds, turnovers, 2-point field goals, touches (touches_min), offensive rebounds, blocks (BLK_mean), assists and fouls committed. The focal case is the forward with higher points scored per minute of play (Kevin Durant). Circles represent training cases and triangles holdout cases. The higher-scoring players have dark filling and the lower-scoring players light filling.

Exercise

Accessing basketball performance requires using holistic perspectives and, therefore, several questions can be explored at the level of critical thinking and decision-making, such as: (1) using variables that provide information from several different dimensions (tactical, technical, physical, …); (2) using variables that isolate individual performance and variables that are related to collective behavior; (3) using variables that describe the performed actions, but also variables or procedures that explore interactions; (4) using variables that describe game macro-structures

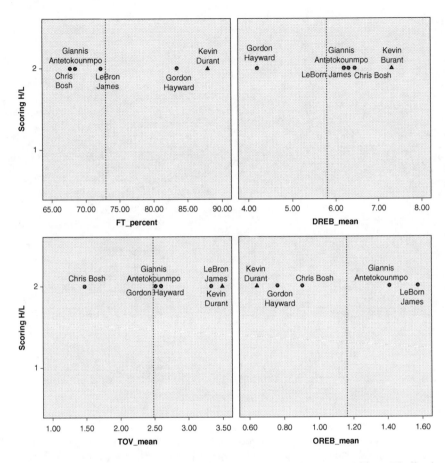

Figure 6.7 Target values by predictors for initial focal cases and nearest neighbors. The focal case is the guard with higher points scored per minute of play (Kevin Durant). Circles represent training cases and triangles holdout cases. The higher-scoring players were allocated with a score of 2 and lower-scoring players with a score of 1 (dummy variable).

(5x5) and micro-structures (1x1, ...); (5) using combined static, dynamic and self-organized perspectives of complexity; and (6) using combined statistical techniques.

Bearing these questions in mind, the tasks can be:

(1) Design a model of basketball performance by describing different levels of analysis and the situational variables that might influence the coupling subject/team-task-environment at each level.
(2) Reinforce the proposed model by designing an adequate evaluation method for each level.

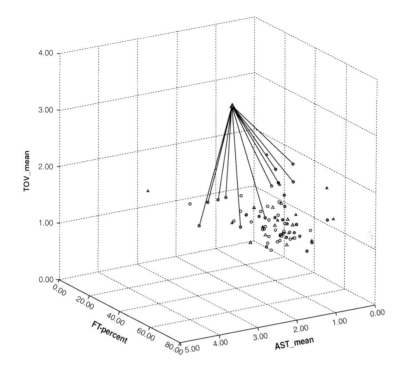

Figure 6.8 Lower-dimensional projection of the predictor space containing seven predictors: free throws (FT_percent), touches, turnovers (TOV_mean), 3-point field goals, assists (AST_mean), defensive rebounds (DREB_mean) and steals. The focal case is the center with higher points scored per minute of play (DeMarcus Cousins). Circles represent training cases and triangles holdout cases. The higher-scoring players have dark filling and the lower-scoring players light filling.

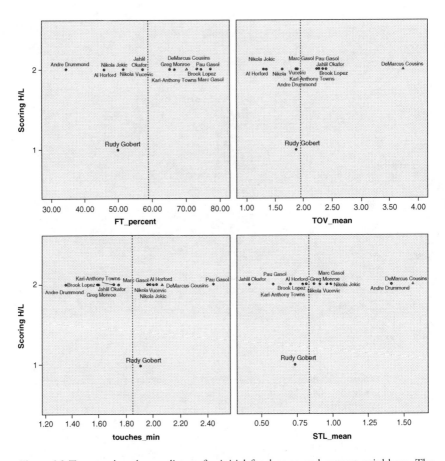

Figure 6.9 Target values by predictors for initial focal cases and nearest neighbors. The focal case is the guard with higher points scored per minute of play (DeMarcus Cousins). Circles represent training cases and triangles holdout cases. The higher-scoring players were allocated with a score of 2 and lower-scoring players with a score of 1 (dummy variable).

References

Arya, S. & Mount, D. (1993). *Algorithms for fast vector quantization*. In J. Storer & M. Cohn (Eds.), Data Compression Conference (pp. 381–390). Los Alamitos: IEEE Computer Society Press.

Ben Abdelkrim, N., El Fazaa, S. & El Ati, J. (2007). Time-motion analysis and physiological data of elite under-19-year-old basketball players during competition. *British Journal of Sports Medicine, 41*, 69–75, discussion 75.

Bracewell, P. (2003). Monitoring meaningful rugby ratings. *Journal of Sports Sciences, 21*, 611–620.

Calleja-Gonzalez, J., Mielgo-Ayuso, J., Lekue, J., Leibar, X., Erauzkin, J., Jukic, I., ... Terrados, N. (2016). The Spanish "Century XXI" academy for developing elite level basketballers: Design, monitoring and training methodologies. *Physician and Sportsmedicine*, *44*, 148–157.

Caparrós, T., Casals, M., Peña, J., Alentorn-Geli, E., Samuelsson, K., Solana, Á., ... Gabbett, T. (2017). The use of external workload to quantify injury risk during professional male basketball games. *Journal of Sports Science and Medicine*, *16*, 480–488.

Chiu, T., Fang, D., Chen, J., Wang, Y. & Jeris, C. (Eds.). (2001) *A Robust and scalable clustering algorithm for mixed type attributes in large database environment*. San Francisco, CA: ACM.

Cooley, W. & Lohnes, P. (Eds.). (1971). *Multivariate data analysis*. New York: John Wiley & Sons, Inc.

Cunningham, P. & Delaney, S. (Eds.). (2007). *k-Nearest neighbor classifiers*. Dublin: University College Dublin.

FIBA (Eds.). (2016). *Basketball statisticians manual*. Geneva: International Basketball Federation.

Friedman, J., Bentley, J. & Finkel, R. (1977). An algorithm for finding best matches in logarithm expected time. *ACM Transactions on Mathematical Software*, *3*, 209–226.

García, J., Ibáñez, S., Martinez De Santos, R., Leite, N. & Sampaio, J. (2013). Identifying basketball performance indicators in regular season and playoff games. *Journal of Human Kinetics*, *36*, 161–168.

Gomez, M., Lago-Peñas, C. & Pollard, R. (2013a). Situational variables. In T. Mcgarry, P. O'Donoghue & J. Sampaio (Ed.), *Routledge handbook of sports performance analysis*. London: Routledge, Taylor & Francis.

Gomez, M., Lorenzo, A., Ibáñez, S. & Sampaio, J. (2013b). Ball possession effectiveness in men's and women's elite basketball according to situational variables in different game periods. *Journal of Sports Sciences*, *31*, 1578–1587.

Gomez, M., Lorenzo, A., Sampaio, J., Ibáñez, S. & Ortega, E. (2008). Game-related statistics that discriminated winning and losing teams from the Spanish men's professional basketball teams. *Collegium Antropologicum*, *32*, 451–456.

Gonzalez, A., Hoffman, J., Rogowski, J., Burgos, W., Manalo, E., Weise, ... Stout, J. (2013). Performance changes in NBA basketball players vary in starters vs. nonstarters over a competitive season. *Journal of Strength & Conditioning Research*, *27*, 611–615.

Ibáñez, S., Sampaio, J., Feu, S., Lorenzo, A., Gomez, M. & Ortega, E. (2008). Basketball game-related statistics that discriminate between teams' season-long success. *European Journal of Sport Science*, *8*, 369–372.

Ittenbach, R., Kloos, E. & Etheridge, J. (1992). Team performance and national polls: The 1990–91 NCAA Division I basketball season. *Perceptual and Motor Skills*, *74*, 707–710.

Karipidis, A., Fotinakis, P., Taxildaris, K. & Fatouros, J. (2001). Factors characterizing a successful performance in basketball. *Journal of Human Movement Studies*, *41*, 385–397.

Kozar, B., Vaughn, R., Whitfield, K., Lord, R. & Dye, B. (1994). Importance of free-throws at various stages of basketball games. *Perceptual and Motor Skills*, *78*, 243–248.

Kubatko, J., Oliver, D., Pelton, K. & Rosenbaum, D. (2007). A starting point for analyzing basketball statistics. *Journal of Quantitative Analysis in Sports*, *3*, 1–22.

Leite, N., Leser, R., Goncalves, B., Calleja-Gonzalez, J., Baca, A. & Sampaio, J. (2014). Effect of defensive pressure on movement behaviour during an under-18 basketball game. *International Journal of Sports Medicine*, *35*, 743–748.

Malarranha, J., Figueira, B., Leite, N. & Sampaio, J. (2013). Dynamic modeling of performance in basketball. *International Journal of Performance Analysis in Sport, 13,* 377–386.

Mateus, N., Goncalves, B., Abade, E., Liu, H., Torres-Ronda, L., Leite, N. & Sampaio, J. (2015). Game-to-game variability of technical and physical performance in NBA players. *International Journal of Performance Analysis in Sport, 15,* 764–776.

Mertz, J., Hoover, D., Burke, J., Bellar, D., Jones, L., Leitzelar, B. & Judge, L. (2016). Ranking the greatest NBA players: A sport metrics analysis. *International Journal of Performance Analysis in Sport, 16,* 737–759.

Oliver, D. (Ed.). (2004). *Basketball on paper: Rules and tools for performance analysis.* Washington, DC: Brassey's, Inc.

Pedhazur, E. (Ed.). (1982). *Multiple regression in behavioral research.* New York: Holt, Rinehart & Winston.

Pojskic, H., Separovic, V., Uzicanin, E., Muratovic, M. & Mackovic, S. (2015). Positional role differences in the aerobic and anaerobic power of elite basketball players. *Journal of Human Kinetics, 49,* 219–227.

Sampaio, J., Gonçalves, B., Rentero, L., Abrantes, C. & Leite, N. (2014). Exploring how basketball players' tactical performances can be affected by activity workload. *Science & Sports, 29,* 23–30.

Sampaio, J., Ibáñez, S. & Lorenzo, A. (2013). Basketball. In T. Mcgarry, P. O'Donoghue & J. Sampaio (Ed.), *Routledge handbook of sports performance analysis.* London: Routledge, Taylor & Francis.

Sampaio, J. & Janeira, M. (2003). Statistical analyses of basketball team performance: Understanding teams' wins and losses according to a different index of ball possessions. *International Journal of Performance Analysis in Sport, 3,* 40–49.

Sampaio, J., Lago, C., Casais, L. & Leite, N. (2010). Effects of starting score-line, game location, and quality of opposition in basketball quarter score. *European Journal of Sport Science, 10*(6), 391–396.

Sampaio, J., Mcgarry, T., Calleja-Gonzalez, J., Saiz, S., Del Alcazar, X. & Balciunas, M. (2015). Exploring game performance in the National Basketball Association using player tracking data. *PLoS One, 10,* e0132894.

Shaoliang, Z., Lorenzo, A., Gómez, M., Liu, H., Gonçalves, B. & Sampaio, J. (2017). Players' technical and physical performance profiles and game-to-game variation in NBA. *International Journal of Performance Analysis in Sport, 17*(4), 466–483.

Tatsuoka, M. (Ed.). (1971). *Multivariate analysis.* New York: John Wiley & Sons, Inc.

Trninic, S., Perica, A. & Dizdar, D. (1999). Set of criteria for the actual quality evaluation of elite basketball players. *Collegium Antropologicum, 23,* 707–721.

Zhang, T., Ramakrishnon, R. & Livny, M. (Eds.). (1996). BIRCH: An efficient data clustering method for very large databases. In Proceedings of The ACM SIGMOD Conference on Management of Data, Montreal, Canada: ACM.

Part III
Evaluation concepts and techniques

7 Tournaments

Peter O'Donoghue

Introduction

Sports tournaments can be organized as round-robin leagues, knockout competitions or a combination of both. There may be qualifying stages required due to restrictions on the number of participating teams or players. Tournament structures have changed to improve participation, revenues, entertainment value and safety. For example, the European nations' soccer tournament increased the number of participating teams from 8 to 16 teams in 1996 and to 24 teams in 2016. Euro 2020 will have 24 teams just like Euro 2016 did. However, there will be no host nation and qualification for the tournament will also change with four of the teams qualifying through play-offs between teams decided by the new UEFA (Union of European Football Associations) Nations League tournament.

The effectiveness of any changes to a tournament can be measured using performance indicators that reflect the areas that organizers intend to improve. These indicators could include the percentage of teams of given rankings qualifying for tournaments and the percentage of these teams reaching the latter stages of tournaments.

Tournament designers can consider whether they have seeding or not, play matches at neutral venues or have host nations, apply alternative match formats and limit squad sizes. In order to make such decisions, it is necessary to have information about the impact of such decisions on the performance indicators relating to entertainment, revenue, participation and safety.

The chapter covers the problem of modelling and simulating sports tournaments to provide decision support information to tournament designers. The use of regression analysis to produce the underlying models used in tournament simulators is described in detail. Regression analyses have assumptions that statisticians argue should be met by data used to form regression models. This chapter argues that sports performance data that violate such assumptions represent important characteristics of the sport that should be modeled. A process is devised for transforming models so that they satisfy the assumptions of modelling techniques; this involves inverse transformations being done when predictions are being made.

The chapter then discusses the use of simulation packages for tournaments and the main algorithms and data structures that are used to produce the output information required by decision makers. This is done using the example of the Euro 2020 tournament with the UEFA Nations League involved in the qualifying process.

The tournament design problem

In order to acquire the information about alternative tournament structures, experimental, analytical or simulation studies can be done. The experimental approach has a serious disadvantage which is that performers need to play in experimental tournaments where tournament structures are untried and untested. This raises practical and ethical concerns. However, tournaments have been played under experimental laws allowing the impact of these to be studied before they are implemented throughout the given sport (Fuller, Raftery, Readhead, Targett & Molloy, 2009; Van Den Berg & Malan, 2012). Analytical approaches require mathematical ability to model tournaments in terms of relevant variables, understand distributions of these variables and prove the impact of these on performance indicators about tournament quality and effectiveness. These models can become very complicated to solve even when a modest set of variables are included. Examples of models being solved by analytical means have been used in tennis research to determine the probability of winning tennis matches for given point probabilities (Fischer, 1980) and the impact of different scoring systems (Carter Jr. & Crews, 1974; Pollard, 1983). Simulation studies allow more complicated models of tournaments to be simulated to provide output statistics for the relevant performance indicators. Computational power available today makes this a flexible approach allowing ad hoc queries about variations of tournament structures to be tested. Once the simulation results are produced and considered by decision makers, the given sports organization is in a much better position to try tournaments with experimental rules and/or structures.

A simulation package is only as good as the underlying model of sports performance upon which it is based. Regression models of expected results can be determined using relevant factors. This requires an analysis of historical cases under current regulations. A key piece of information that is required to produce a simulation system is the distribution of the residual values for match outcomes. This distribution of the residual values allows simulated tournaments to include random variation about the expected results that is evidenced in historical cases. Simulation systems can then simulate tournaments thousands of times, accumulating progression statistics for different types of situation. For example, Newton and Aslam (2006) compared the chances of winning tennis matches when "hot hand effects" were present and absent in simulated performances. This validated models of tennis performance based on the assumptions of stationarity and independence of points.

There are some aspects of tournaments that are easier to quantify than others; we have already mentioned the chances of teams of different qualities qualifying for tournaments and reaching different stages of the tournaments. The number of matches played in a tournament can be determined for each team as well as for the tournament as a whole. Outcomes such as improving entertainment value are more difficult to quantify.

Tournaments can be studied by simulating performances within the same structures of matches used in the actual tournaments. These structures may be round-robin tournaments, knockout tournaments or a combination of both. For example, simulations of Gaelic football (O'Donoghue, 2005), soccer (O'Donoghue, 2014), rugby (O'Donoghue, Ball, Eustace, McFarlan & Nisotaki, 2016), netball (O'Donoghue, 2012b) and tennis (O'Donoghue, 2013) tournaments have "played" every match within a simulated tournament. These tournament simulations rely on realistic simulation of individual performances with respect to the information required by decision makers. In some studies, whole matches are simulated at a high level of abstraction without individual situations within the matches being simulated (O'Donoghue, 2014; O'Donoghue et al., 2016). There are other studies where internal match events are simulated; for example, there have been simulations of tennis tournaments where each point within each match has been simulated (Newton & Aslam, 2006).

The problem area of interest to the current chapter is modelling European international soccer performance so that simulations can be used to compare alternative structures for European international soccer tournaments. The key outcome variables of interest are the chances of teams of different qualities reaching different stages of the tournaments.

Concepts and methods of modelling

Individual sports performances can be modeled using discrete or continuous probability distributions of outcomes. A discrete probability distribution classifies outcomes broadly (for example, win, draw or loss) and requires probabilities to be assigned to the outcomes. A continuous probability distribution is used where the outcome variable is a time, distance or some other continuous variable. In this case, the probability generated by a random number generator represents the probability of the time or distance being a particular value or less. However, for the purposes of simulation, this probability is used to select a single value for the outcome variable of interest which is the value at the upper end of the range of values represented by the given probability. Continuous probability distributions have also been used where outcomes variables are integers rather than real numbers; for example, the winning margin in a basketball match. Here, negative values represent losses of different margins, positive values represent wins of different margins and values around zero represent a draw. A random probability is associated with a real number for the outcome variable because the probability distribution is on a continuous scale. The real number is rounded, for example values of between -0.5 and +0.5 round to 0, representing a draw, while values between 11.5 and 12.5 round to 12 representing a win by 12 points.

The probability distribution of a match result comes from an expected result using a regression model and the known distribution of residual values for the model. Effectively, we are using the expected result to reflect the chance of higher-quality teams winning the match but using the residuals to reflect variability about this expected result. The data used to create regression models need to satisfy the assumptions of the modelling technique used. For example, in linear regression, residual values should be normally distributed (Vincent, 1999: 111), homoscedastic (Anderson, Sweeney & Williams, 1994: 521) and independent (O'Donoghue, 2012a: 151). However, this chapter argues that data violating such assumptions provide an opportunity for modelers. For example, if residuals show a shotgun effect (heteroscedasticity) then this can be built into underlying models used by simulation systems. Similarly, if there are ordering effects in residual values, these trends can be included within models. Independent variables should not be highly correlated with each other (O'Donoghue, 2012a: 161). Further assumptions are that there should be at least 20 cases for each independent variable (Ntoumanis, 2001: 120–121), there should be no outliers or extreme values in the independent or dependent variables (Tabachnick & Fidell, 2007: 124) and there must be no multivariate outliers (Ntoumanis, 2001: 124–125).

Methodological experience

Modelling phase

The current section proposes an approach to performance modelling that provides the necessary parameters for a simulation system. The model is comprised of a regression equation for the expected result and the distribution of possible results about this expected result. The regression equation is determined by analyzing previous matches of known outcomes, as illustrated in Figure 7.1. The regression equation is created using curve fitting and selecting the most significant of the alternative regression lines. For example, linear, inverse, logarithmic, quadratic and cubic regression models could all be attempted. Quadratic and cubic models should be used with caution as the curves generated have one and two turning points respectively. Quadratic models are valid where there is a uniform rate of change in a dependent variable. Human behavior is often more dynamic and complex than this. This could mean that while quadratic or cubic models may be more significant than some others, the pattern of predicted outcome values might not be logical with respect to the sport. For example, the quadratic model could show the outcome variable increasing up to a point when plotted against some independent variable and then decreasing beyond that point. Conceptually, the outcome variable might be expected to increase as the independent variable increases. If the turning point in the relationship between the two variables is within the meaningful range of values for the independent variable, then the quadratic model should not be used; otherwise, it could be used.

The regression equation is the best straight- or curved-line model that can be fitted to the data from previous matches. As illustrated in Figure 7.1, few (if any)

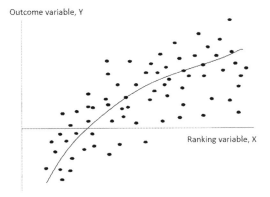

Figure 7.1 Model using regression analysis.

points are actually on the line. In this example, the outcome variable Y is modeled in terms of a ranking variable X. For any given match in the data set used to create the model, the predicted value of Y according to the model is termed Y'. The difference Y–Y' is referred to as a residual value, E, representing the difference between the observed and predicted values.

The data used to create the models should satisfy the assumptions of the regression technique used. These assumptions were listed in the previous section. Normality can be tested using a Kolmogorov Smirnov test, which has a greater strength when there are at least 50 previous matches in the data set, or a Shapiro Wilk test if there are fewer than 50 previous matches used. Where the data are not normally distributed, the nature of the violation of the assumption can be considered using z_{Skew} and z_{Kurt} for skewness and kurtosis respectively. The residual values need to be homoscedastic, meaning that the distribution of the residual values should be similar no matter the values of X and Y'. Figure 7.2 shows an example of where this assumption is violated; the data are heteroscedastic because there is a "shotgun effect" with the residual values showing a larger variance for higher values of X and Y' than for lower values. Rather than relying on visual inspection of a scatter plot of previous cases, homoscedasticity can be assessed more rigorously by determining the correlation between predicted values Y' and absolute residual values (removing the minus sign from any negative residual values). If this correlation coefficient is less than 0.25, then the data can be considered homoscedastic (O'Donoghue, 2012a: 357–358). A final assumption is that there should be no order effect in the residuals. This means that the spread of residual values should be similar irrespective of the historical order in which the previous matches were played. There are a number of ways that this assumption can be assessed. One way is to correlate residual values with the date of the match. Alternatively, previous matches can be grouped into the tournaments where they occurred and the

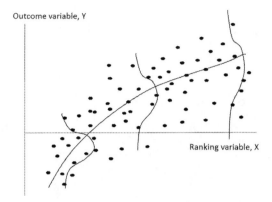

Figure 7.2 An example of heteroscedasticity within data (the vertical lines indicate the distribution of residual values for given values of the ranking variable).

residual values can be compared between different previous tournaments using an ANOVA test.

As mentioned in the introduction to this chapter, the approach proposed here is not to force data to satisfy the assumptions of the modelling technique used. The reason why normality of residual values is important is because the normal distribution may be used during simulation. Serious violations of this assumption in previous case data could render the use of the normal distribution inappropriate during simulation. Therefore, alternative distributions may need to be programmed into simulators or residual values may need to be mapped onto the standard normal distribution. This would allow the normal distribution to be used to simulate random variability about residual results with inverse mapping being used to ensure that simulated random variation follows the correct distribution of residual values. Alternatively, a Monte Carlo simulation could use the observed distribution of residual values.

Homoscedasticity of residuals is important where simulation systems apply random variation using the same distribution of residual values irrespective of the expected value Y' of the outcome variable. The current approach views the violation of this assumption as a modelling opportunity. The approach deals with violations of the assumption of homoscedasticity by programming simulators to use different distributions of residuals for different expected outcome values. These different distributions are evidenced by the differing spreads of residual values in historical cases.

Residual values may also have order effects that violate the final assumption of regression analysis; this is also an opportunity to be exploited during simulation studies of tournaments. There may be tournaments where the strength in depth is gradually increasing over time with increasing numbers of upsets being observed. Such trends can be extrapolated to future tournaments that are being simulated by adjusting random variability as evidenced by the trends.

Simulations phase

Matches between any given pair of opponents will not always lead to the same outcome. Therefore, simulating a match or an event within a match involves generating a result which is one of many possible outcomes. This requires the "throwing of a dice" or generating a random number to select the particular outcome to be produced by the simulator. This throwing of the dice is biased by the expected chance of given results. For example, if previous evidence suggest that Team A has a probability of beating Team B of 1/2, a probability of 1/3 of drawing with Team B and a probability of 1/6 of losing to Team B, then we can roll a 6-sided dice with 4, 5 or 6 representing a win for Team A, 2 or 3 representing a draw and 1 representing a win for Team B. This could also be accomplished within a Monte Carlo simulation. Some simulations run tournaments tens of thousands of times, simulating each match within each repetition of the tournament. This is done in order to generate higher-level tournament-level probabilities of different teams progressing to different stages of the tournament. Researchers would not wish to manually throw a dice thousands of times during the study and a 6-sided dice does not allow event probabilities to be expressed more finely than multiples of 1/6. Therefore, computers are used to generate random numbers on a more continuous scale.

Results

This section describes the resulting model of European international soccer matches. The model needed to cater for matches played at neutral venues as well as matches where one of the teams may be at home. The model of European international soccer performance was created using 616 matches played in previous tournaments since mid-2008: the 31 matches of Euro 2008, 239 matches from the qualifying tournament for Euro 2012, the 31 matches of Euro 2012, 248 matches from the qualifying tournament for Euro 2016 and the 51 matches of the Euro 2016 tournament. The 10 matches of the 2016 qualifying tournament that involved Gibraltar were not included because Gibraltar did not have a UEFA national team ranking coefficient during the qualifying campaign.

Each match was represented by three variables that were expressed as differences between the higher- and lower-ranked team within each match according to the UEFA national team coefficient ranking. The first variable is the difference between the two teams' ranking points at the time the match was played. The second variable represents the match venue. If the higher-ranked team is playing at home, then the value is +1; if the lower-ranked of the two teams is playing at home, then the value is -1; if the match is being played at a neutral venue, then the value is 0. The values of +1 and -1 are used in both qualifying tournaments as well as in matches where the host nation is the higher-ranked and the lower-ranked of the two teams within a match respectively.

The result of the match was represented by how many more goals the higher-ranked team within the match scored than the lower-ranked team. Therefore, values of 0 represent drawn matches while positive values represent matches won by the higher-ranked teams within the matches and negative values represent upsets.

The relationship between UEFA national team coefficient ranking points and match result was explored using the curve-fitting regression analysis feature of SPSS version 22 (SPSS Inc., an IBM company, Armonk, NJ). Regression equations were determined for linear, inverse, quadratic and cubic models. All of these models were highly significant ($p < 0.001$) except for the inverse model ($p = 0.400$). Kolmogorov Smirnov tests found that the assumption of normality of residual values was violated more for the quadratic ($p = 0.004$) and cubic models ($p = 0.015$) than for the linear model ($p = 0.042$). The absolute residual values were not highly correlated with predicted values for goal difference for the linear ($r = +0.149$), quadratic ($r = +0.130$) or cubic models ($r = +0.132$). A series of one-way ANOVA tests was used to test if the residual values differed between the five tournaments from which the previous cases were drawn. These tests revealed no significant order effect on the residual values for the linear ($p = 0.300$), quadratic ($p = 0.223$) or cubic models ($p = 0.217$). Given the satisfaction of the assumptions of homoscedasticity of residual values and no order effect of residual values, as well as the violation of normality being mildest for the linear model, it was decided to test the linear model when the Venue variable was added.

The linear regression was done entering UEFA national team coefficient ranking points difference (RPDiff) first and then the Venue variable. This found that both variables were significant predictors of goal difference between the teams ($p < 0.001$). The model had an adjusted R^2 value of 0.229 and contained the regression coefficients shown in Equation (1). This shows that playing at home is worth 0.328 goals. The contribution of UEFA national team coefficient ranking can be illustrated by considering the match between France and Albania in Euro 2016. These teams had 30,992 and 19,151 UEFA national team coefficient ranking points respectively at the time of the match. The difference of 11,841 ranking points weighted by the regression coefficient of 0.000132 reveals that the difference between the two teams' ranking points was worth 1.563 goals in favor of France.

$$\text{Predicted GD} = -0.263 + 0.000132\ \text{RPDiff} + 0.328\ \text{Venue} \qquad (1)$$

The residual values from this model have a standard deviation of 1.605 but are not quite normally distributed according to a Kolmogorov Smirnov test ($p = 0.030$). This was due to one extreme value (Holland 11–0 San Marino) and 11 outliers. Six of the outliers were where higher-ranked teams won by more goals than expected while the other five outliers were upsets (Slovakia 0–4 Armenia; Holland 0–3 Turkey; Greece 0–1 Faroe Islands; Holland 0–2 Iceland; and Greece 0–2 Northern Ireland). These are results of real matches rather than being down to measurement error. It was, therefore, decided not to remove the outliers or extreme values. An attempt was made to map the residuals onto a

Table 7.1 Standard deviation of absolute residual values.

Predicted GD (rounded)	Matches	Mean absolute residual value
−1	3	0.8349
0	205	1.1047
1	247	1.1721
2	120	1.3464
3	37	1.6918
4	4	2.7338

standard normal distribution. However, the constant of the resulting mapping function was negligible (-2.312E-16), meaning that the mapping equation was merely a scaling function and that the transformed residuals would violate the assumption of normality to the same extent as the untransformed values. It was, therefore, decided not to transform the residual values.

There was very mild heteroscedasticity in the residuals ($r = 0.167$) which did not violate the assumption of homoscedasticity ($r < 0.250$). However, inspection of mean absolute residual values does reveal different spreads of residual values for different predicted goal difference values, as shown in Table 7.1.

Given the steady increase in the spread of random variation about the expected result as predicted goal differences increase, it was decided to incorporate this into the simulator. Linear regression determined Equation (2) for absolute residual value in terms of predicted goal difference:

$$\text{Absolute Residual} = 1.028 + 0.196 \text{ Predicted GD} \qquad (2)$$

However, the simulator needs to know the standard deviation of residuals for any given predicted goal difference. The mean absolute value of 1000 evenly spread values of the standard normal distribution (using probabilities of 0.0005 to 0.9995 in steps of 0.001) is 0.7977. Given that the standard deviation of the standard normal distribution is 1, we need to scale the mean absolute residual up by a factor of $1/0.7977$ in order to generate the standard deviation of residuals the simulator should be applying for any predicted goal difference value. Equation (3) is used to determine the standard deviation to be applied by the simulator.

$$SD = (1.028 + 0.196 \text{ Predicted GD})/0.7977 \qquad (3)$$

The residual values from the revised model including RPDiff and Venue were compared between Euro 2008, Euro 2012, Euro 2016 and the qualifying tournaments for Euro 2012 and Euro 2016. This did not reveal any order effect in the residual values ($p = 0.235$), meaning that the assumption of independence of residuals was satisfied. A limitation of the model is that it is a very simple model with a low R^2 value. Other variables such as points standings within pools at the start of matches and number of matches remaining could be added to address situations where teams may play apply different tactics if they have already qualified from a pool or there is no possibility to qualify from the pool.

Example: Euro 2020 and the UEFA Nations League

The tournaments

The motivation for changing the qualifying processes for Euro 2020 and the introduction of the UEFA Nations League include rejuvenating and improving the quality and standing of national team soccer (www.uefa.com/uefaeuro/news/newsid=2079553.html accessed June 3, 2017). UEFA also wishes to maintain the balance between club and international soccer (www.uefa.org/about-uefa/executive-committee/news/newsid=2191264.html accessed November 3, 2017). These outcomes are difficult to quantify, but some other benefits are more obvious. For example, the increase in competitive fixtures and reduction in friendly matches are determined by the tournament structures when considered together with the constraints of the international calendar used in soccer. Other benefits claimed by UEFA are additional ways for middle- and lower-ranking nations to qualify for Euro 2020. The chances of such teams qualifying for tournaments under the current and proposed tournament structures is something that can be evaluated using simulation.

There are three tournaments of interest in the current example: the UEFA Nations League, the qualifying tournament for Euro 2020 and the Euro 2020 tournament itself. It was not possible to create separate models for these three tournaments because the European Nations League has never been played before. Therefore, the single model of European international soccer performance described in the previous section was used for all three competitions. Like the Euro 2016 tournament, the Euro 2020 tournament will have 6 round-robin pools of 4 teams with 16 teams qualifying for the second round. The knockout tournament includes the second round, quarter-finals, semi-finals and the final, meaning that there are 51 matches in the tournament. One aspect of the tournament that will differ to Euro 2016 is that there is no single host nation for Euro 2020 and the matches will be played in 13 different countries.

There will be a qualifying tournament of round-robin groups with teams playing each other at home and away, as before. However, only 20 of the 24 teams participating in Euro 2020 will qualify from this qualifying tournament. The qualifying tournament will consist of ten groups (five of which have five teams and the other five have six teams) with the group winners and runners-up qualifying for Euro 2020. The other difference to before is that all 55 European national teams will compete in the qualifying tournament because there is no automatically qualifying host nation.

The remaining four places in Euro 2020 are decided by four sets of play-off matches with each set involving four teams who play semi-finals and a final. The 16 teams participating in these play-off tournaments are decided based on the UEFA Nations League. The UEFA Nations League is a new tournament that contains

Tournaments 139

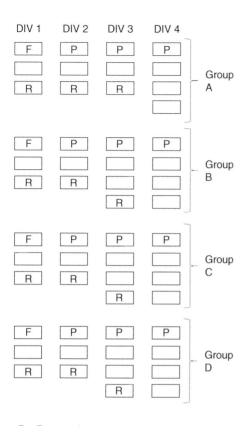

P = Promoted
R = Relegated
F = Play in Final-Four Tournament

Figure 7.3 Structure of the UEFA Nations League.

4 divisions of 12 to 16 teams, as shown in Figure 7.3 (www.uefa.com/uefaeuro-2020/news/newsid=2437695.html accessed November 3, 2017). Each division consists of four groups. Teams will initially be placed into the four divisions of the UEFA Nations League using UEFA national team coefficient rankings published on November 15, 2017 once the qualifying tournament for the 2018 FIFA World Cup has completed. The UEFA Nations League matches are played between September and November 2018 with teams in the same groups playing each other at home and away. The four group winners in the top division contest a final-four tournament which will be played in June 2019. This final-four tournament will consist of two semi-finals that are drawn without considering points gained within groups or UEFA national team coefficient ranking, and a final. The winners of the four groups within the other divisions are promoted to the division above, replacing four teams that are relegated from those divisions. The UEFA

Nations League takes place once every 2 years in between Euro and World Cup tournaments or vice versa.

The main qualifying tournament for Euro 2020 takes place between March and November 2019 with teams in the same groups playing each other at home and away. Once the 20 qualifiers from the 10 qualifying groups are decided, the 16 teams to contest the play-off matches are decided from the UEFA Nations League standings. These 16 teams will be teams that have not already qualified for Euro 2020 with four teams coming from each division decided as follows: Any group winners within a division, who have not already qualified for Euro 2020, participate in the play-off tournament. Where the group winners do not provide four teams for the play-offs, the remaining teams will be the higher-ranked teams within the division that have not already qualified. If necessary, teams from a lower division will fill the remaining places if fewer than four teams from the division have not already qualified from the main qualifying tournament. The play-offs will be played in March 2020 with each division's play-offs involving two semi-finals and a final.

Simulation

Figure 7.4 shows the structure of the simulator that was used to study the campaign comprising the UEFA Nations League, the Euro 2020 qualifying

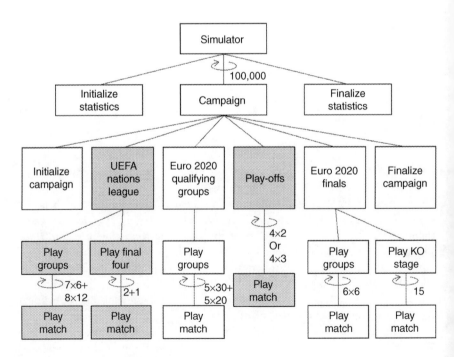

Figure 7.4 Simulator design.

tournament and the Euro 2020 finals. The simulator was programmed in Matlab to read in ranking data for all European soccer nations and then simulate matches within the tournament structures of interest. Matches were simulated by generating a random number between 0 and 1 to represent a probability that was then applied to the distribution of residual matches to determine how many more or fewer goals a team won a match by than predicted. The simulator ran the campaign 100,000 times so that progression statistics could be accumulated and reported. A second version of the simulator was created to simulate the campaign, based on the previous format used for Euro 2016. This second simulator did not include the UEFA Nations League and the play-offs were contested between the eight best-performing third-placed teams in the qualifying groups. The key component used in both simulators was the PlayMatch subroutine which used the model shown in equations (4.1) and (4.3). The UEFA national team coefficient ranking points for each team (on December 3, 2017) is used to determine the difference between the ranking points of the higher- and lower-ranked of the two teams. If the higher-ranked team is playing at home, then Venue is set to +1; if the lower-ranked team is playing at home, then Venue is set to -1; otherwise, Venue is set to 0. Equation (4.1) is then used to determine a predicted goal difference between the teams. If the same two teams play each other 100,000 times, the result will not always be the same. The simulator therefore modifies the predicted result using the random variation evidenced during the analysis of previous tournaments. Equation (4.3) is applied to the predicted goal difference to determine the standard deviation of the normal distribution to be used to generate a random result. A random probability is generated and mapped onto the corresponding value of a normal distribution with a mean being the predicted result and the standard deviation computed in Equation (4.3). This is the simulated result of the match in the given campaign; remember there are 100,000 campaigns run by the simulator. In pool matches, simulated goal differences of between -0.5 and +0.5 are draws, values greater than +0.5 represent wins for the higher-ranked of the two teams and values less than -0.5 represent upsets. In knockout matches, values of greater than or equal to 0 are wins for the higher-ranked team and values of less than 0 are upsets. This is because one team will progress and the other will be eliminated even if extra time and penalties are required.

Table 7.2 shows the percentage of simulated tournaments where different teams were promoted, relegated and won the UEFA Nations League 2018–2019. This suggests that Germany is the favorite to win the tournament, but there is a 71% chance that the tournament will be won by a team other than Germany.

Table 7.3 shows the results of the simulation study comparing the two campaign structures. There are 24 teams that will qualify for the Euro 2020 tournament. Table 7.3 shows the mean number of teams from each Nations League division that qualify for and reach each stage of the tournament under the two qualifying systems within the simulated tournaments. As we can see, the number of Division 1, 2, 3 and 4 teams qualifying for Euro tournaments and progressing to different stages is very similar between the two tournaments.

Table 7.2 Chances of team success in the UEFA Nations League 2018–2019.

Division	Team	Percentage simulation runs with promotion	Percentage simulation runs with relegation	Win UEFA Nations League
1	Germany		7.6	29.0
	Spain		15.2	17.5
	England		24.4	9.9
	Portugal		29.1	8.0
	Belgium		29.6	6.6
	Italy		29.8	7.0
	Netherlands		33.9	5.7
	France		34.2	6.0
	Russia		47.0	2.7
	Switzerland		47.8	2.7
	Austria		49.7	2.5
	Croatia		51.7	2.4
2	Bosnia and Herzegovina	40.1	26.8	
	Ukraine	39.9	26.9	
	Czech Republic	34.2	32.4	
	Sweden	31.6	35.0	
	Poland	33.1	33.7	
	Romania	31.6	35.2	
	Slovakia	26.1	41.1	
	Hungary	26.2	41.3	
	Denmark	35.6	30.6	
	Turkey	35.1	31.4	
	Republic of Ireland	34.0	32.0	
	Greece	32.6	33.7	
3	Norway	33.5	30.3	
	Slovenia	31.2	31.6	
	Iceland	36.9	26.1	
	Wales	36.1	26.1	
	Israel	30.9	18.6	
	Scotland	25.9	22.6	
	Albania	27.3	22.0	
	Montenegro	26.7	22.1	
	Northern Ireland	31.7	18.2	
	Serbia	27.2	22.2	
	Finland	22.4	26.9	
	Bulgaria	21.3	28.3	
	Armenia	13.5	38.8	
	Estonia	16.8	34.5	
	Lithuania	18.7	31.7	
4	Belarus	51.4		
	Georgia	44.9		
	Azerbaijan	38.3		
	Latvia	38.1		
	Cyprus	43.1		
	Moldova	40.7		

Table 7.2 (Cont.)

Division	Team	Percentage simulation runs with promotion	Percentage simulation runs with relegation	Win UEFA Nations League
	Macedonia	32.5		
	Kazakhstan	23.4		
	Luxembourg	23.0		
	Liechtenstein	17.3		
	Faroe Islands	17.0		
	Malta	11.0		
	Kosovo*	8.4		
	Andorra	4.5		
	San Marino	3.9		
	Gibraltar	2.6		

Note: based on rankings published on October 14, 2015
* Kosovo did not have a UEFA national team coefficient ranking when the last rankings were published in October 2015. Based on current FIFA (Fédération Internationale de Football Association) ranking points and their position relative to other European nations, Kosovo was allocated 10,000 points under the UEFA ranking system

Table 7.3 Progression statistics produced by the Euro 2020 simulator.

Condition	Without UEFA Nations Lg				With UEFA Nations Lg			
	Div 1	Div 2	Div 3	Div 4	Div 1	Div 2	Div 3	Div 4
Qualifying tournament								
Qualify from qual group	10.9	6.7	2.3	0.1	10.9	6.7	2.3	0.1
Qualify via play-offs	0.6	1.8	1.6	0.1	0.3	1.3	1.4	1.0
Euro 2020 tournament								
Make Round 2	9.5	5.0	1.4	0.0	9.5	4.9	1.4	0.1
Make quarter-finals	5.5	2.1	0.4	0.0	9.5	4.9	0.4	0.0
Make semi-finals	3.1	0.8	0.1	0.0	3.1	0.8	0.1	0.0
Make final	1.7	0.3	0.0	0.0	1.7	0.3	0.0	0.0
Win Euro 2020	0.9	0.1	0.0	0.0	0.9	0.1	0.0	0.0

The main difference is that the new campaign structure guarantees that at least one Division 4 team will qualify for the tournament through the play-offs. This is at the expense of teams in higher divisions that had higher chances of qualifying through the play-offs under the previous system. Teams from Division 1 of the UEFA Nations League won 90.4% of simulated Euro 2020 tournaments, the most likely winning teams being Germany (24.8%), Spain (15.5%), England (9.2%) and Portugal (7.5%) with other teams winning the remaining 42.9% of the simulated tournaments.

The initial analysis of the Euro 2020 tournament leads to further questions that can be answered by adding statistical monitoring code to the simulators. For example, there may be a query as to whether teams drawn in the same pool as the Division 4 play-off qualifier have an increased chance of qualifying for Round 2 as one of the best four third-placed teams. Code was added to the PlayGroups subroutine of the Euro2020 simulator and the Initialization routine to count the number of occasions where this happened. This revealed that the third-placed team in the group containing the Division 4 play-off winner qualified on 71.2% of simulated campaigns, which is above the 66.7% expected if all groups have an equal chance of their third-placed team progressing (four of the six third-place teams qualify for the knockout stages). However, this increased chance of qualification was not as high as some might have imagined.

Conclusions

Tournament structures develop over time to make changes that are intended to benefit the sport, players and spectators. Simulation has a role in evaluating the effects of proposed changes before they are experimented with and eventually implemented throughout the sport. There is a range of different tournament variables that can be studied when simulators are applied. This chapter has used the introduction of the UEFA Nations League and changes to the qualification process for the Euro 2020 soccer tournament as an example showing the differences in teams' chances of qualifying for the tournament under the proposed, and the previous, systems. The information produced can be considered by decision makers during tournament development when various options may be explored. Simulation does have some limitations based on the models used and the assumptions made. For example, in the case of the UEFA Nations League and Euro 2020, it has been assumed that variability in soccer performance will be influenced in a uniform way within qualifying group matches, the UEFA Nations League, play-offs and the Euro 2020 tournament. It is also assumed that teams will be equally competitive and have similar player availability for all of these different types of match. These limitations need to be taken on board by decision makers when using information produced by simulation of tournaments.

Exercise

Consider a tournament in a sport of your choice and propose a change to the tournament organization. For example, seeding could be introduced, matches could be played over one or two legs or the number of sets could change. Model individual match performance in the sport using previous data and some independent ranking variable representing team or player quality. Use the approach outlined in this chapter to represent variability about predicted performance. Now construct two versions of a simulator to run the tournament at least 1,000 times each and produce statistics for carefully chosen tournament indicators of interest. One simulator should play the tournament in its current form and the other should play the tournament when your proposed rule change is applied.

References

Anderson, D., Sweeney, D. & Williams, T. (1994). *Introduction to statistics: Concepts and applications* (3rd edn.). Minneapolis/St. Paul, MI: West Publishing Company.

Carter Jr., W. & Crews, S. (1974). An analysis of the game of tennis. *The American Statistician*, 28(4), 130–134.

Fischer, G. (1980). Exercise in probability and statistics, or the probability of winning at tennis. *American Journal of Physics*, 48(1), 14–19.

Fuller, C., Raftery, M., Readhead, C., Targett, G. & Molloy, M. (2009). Impact of the International Rugby Board's experimental law variations on the incidence and nature of match injuries in southern hemisphere professional rugby union. *South African Medical Journal*, 99(4), 232–237.

Newton, P. & Aslam, K. (2006). Monte Carlo tennis. *SIAM Review*, 48(4), 722–742.

Ntoumanis, N. (2001). *A step by step guide to SPSS for sport and exercise studies*. London: Routledge.

O'Donoghue, P. (2005). The role of simulation in sports tournament design for game sport. *International Journal of Computer Science in Sport*, 4(2), 14–27.

O'Donoghue, P. (2012a). *Statistics for sport and exercise studies: An introduction*. London: Routledge.

O'Donoghue, P. (2012b). The effect of rule changes in World Series Netball: A simulation study. *International Journal of Performance Analysis in Sport*, 12, 90–100.

O'Donoghue, P. (2013). Rare events in tennis. *International Journal of Performance Analysis in Sport*, 13, 535–552.

O'Donoghue, P. (2014). Factors influencing the accuracy of predictions of the 2014 FIFA World Cup. *International Journal of Computer Science in Sport*, 13(2), 32–49.

O'Donoghue, P., Ball, D., Eustace, J., McFarlan, B. & Nisotaki, M. (2016). Predictive models of the 2015 Rugby World Cup: Accuracy and application. *International Journal of Computer Science in Sport*, 15(1), 37–58.

Pollard, G. (1983). An analysis of classical and tie-breaker tennis. *Australian Journal of Statistics*, 25(3), 496–505.

Tabachnick, B. & Fidell, L. (2007). *Using multivariate statistics* (5th edn.). New York: Harper Collins.

Van Den Berg, P. & Malan, D. (2012). The effect of experimental law variations on the Super 14 Rugby Union tournament. *African Journal for Physical, Health Education, Recreation and Dance*, 18(3), 476–486.

Vincent, W. (1999). *Statistics in kinesiology* (3rd edn.). Champaign, IL: Human Kinetics.

8 Key performance indicators

Jürgen Perl and Daniel Memmert

Introduction

As set out in Perl and Memmert (2017), "The intention of Key Performance Indicators (KPI) is to map complex systems behavior to single numbers for scaling, rating and ranking systems or system components." Obviously, such an attempt is difficult from several points of view.

For instance, take the mean income of the inhabitants of a city, a region or a country as an indicator for its wealth. What does this indicator say, and what can it be used for? The mean income indicator is not really informative regarding the income situation of the people because the mean operator reduces the information about the individual inhabitants to an abstract number regarding an abstract object, i.e., the "mean inhabitant." Moreover, the same indicator value can reflect quite different situations from equal incomes of all inhabitants on the one side to a distribution ranging from very high to very low incomes on the other. Normally, in statistical analyses, at least the problem of different distribution types is tried to be solved by adding the variance of the distribution as an indicator of second order, completing the first-order information of the mean value. Continuing this logic leads to a system of indicators of increasing order, each of which completes the information of the higher levels with decreasing significance. This method is similar to that of approximating arbitrary mathematical functions by a system of standardized functions of a same type, i.e., polynomials or sin-/cos-functions. In this correspondence, the mean income as a single indicator is as appropriate for characterizing the income distribution as a constant linear function is for characterizing the complex behavior of a complicated mathematical function.

But despite the fact that the mean value is not adequately modelling the distribution inside the system, is the mean income value at least helpful for comparing cities, regions or countries regarding their level of wealth, assumed that it represents such wealth anyway?

Yes and no.

Yes, if the indicator value is just used for a syntactical ranking, i.e., higher value implies higher position in the ranking list. Logically, this means a simple tautology.

No, if the indicator value is additionally used for semantic interpretations or implications like sorting in groups of similar quality. For instance, assume a wealth value scale from 0 to 30, where values below 10 mean "poor," values between 10 and 20 mean "good" and values between 20 and 30 mean "excellent." Assume further two cities A and B, the values of which are vA = 25 and vB = 15. Obviously, it makes sense to say that A is of excellent wealth while B is only of good wealth. But what about the case that vA = 20.1 and vB = 19.9, in particular if membership of a quality group is reason for a reward or investment or something else? In this case, the income of a handful of inhabitants could influence the future of the whole city significantly.

Back to the area of sports, the term "performance" replaces the terms "income" and "wealth," and KPIs are used to measure the performance of athletes and teams.

If the focus is on a single athlete, it is comparably easy to define appropriate KPIs, as long as they can be derived from single measurable facts like the speed of a runner or the power of a weightlifter. It becomes significantly more difficult if performance is defined by a set of attributes characterizing components of performance. Examples are the biometric profile of an athlete or the technical and the tactical performance of an athlete, particularly of players in games, where pieces of information have to be aggregated and abstracted to one characteristic number. Finally, those individual KPIs can be combined to player-depending team-KPIs in team sports. Alternatively, the team can be handled as a second-level object with measurable activities like attacks or goals.

The derived KPIs can be used in (at least) two ways. A first way is to follow the time-depending sequences of the KPI values of athletes or teams, e.g., to detect reasons for weak phases or to confirm the success of training concepts. A second way is to compare the KPI values of the members of sets of athletes or teams in order to generate rankings.

The listed aspects and problems are discussed in more detail in Part II with a particular focus on soccer. Modelling approaches are presented and discussed in Part III. A detailed list of references regarding modelling in sports is given in Perl (2015).

Project/problem

The idea of using soccer as a basis for discussing KPIs is based on the fact that it combines most of the aspects of KPIs from the single athlete up to the level of team activities. In particular, the aspect of ranking in combination with the quality of player and team behavior is important, as pointed out in Perl and Memmert (2017):

> Soccer is a complex game with a lot of technical and tactical facets (Rein & Memmert, 2016). Win or loss of a game often depends on small and often either hardly detectable details or complicated processes. Nevertheless, 0, 1 or 3 points for the result of the game assesses the game simply by the result, and the sum of those points at the end of the season is an important indicator, qualifying the team for higher challenges or not. This means

that there is a basic and very simple mapping of complex quality to simple quantity. Insofar it is completely consistent to map qualitative aspects of the game like tactical behaviour to quantitative ratings like percentage of success.

(Perl and Memmert, 2017, p. 66)

Example: soccer

From bottom (player) to top (team), there is a large number of quite simple KPIs in soccer (for a recent overview see Memmert, Lemmink and Sampaio, 2017). Some examples are given in Table 8.1.

All those countable indicators may be used for getting a first impression of what was good or bad. But, as pointed out in the introduction, they fail in giving an idea of better or worse. For example, if the number of played passes of player A is 19 and that of player B is 20: Does this mean that player B is better at passing than player A? Normally, it does not. This has to do with the difference between human thinking and mathematical logic. From a mathematical point of view, 20 is greater than 19. And if the number indicates quality, then B is better than A. In contrast, experience says that, if 20 is good, then 19 has to be at least not bad and not really worse than 20. This discrepancy between human thinking and mathematical logic was one reason for developing Fuzzy modelling, saying that not a fixed interval of attributes belongs to a single property but the property (with different degrees of membership) belongs to the attributes of overlapping intervals, i.e., interval "good" may contain the pass numbers 8 to 22 while the interval "excellent" may contain the pass numbers 18 to 30. The result is that player A and player B are both members of the classes "good" and "excellent" (although to a different extent), so there is no reason for ranking B better than A (also see Figure 8.1). A similar approach can be found in the clustering concept

Table 8.1 Some examples of KPIs on increasing levels of complexity.

Unit	KPIs
Player	maximum speed [km/h]
	running performance [km]
Player, tactical group	number of played passes
	mean length of passes
	mean number of outplayed players per pass
	ball losses
	ball recoveries
Tactical group, team	number of attacks
	number of ball-control events
	space-control rates
Team	number of goals
	result
	ranking

of self-organizing neural networks; both approaches are introduced and discussed in more detail in Part III.

Another problem is that the countable indicators, which are exemplarily introduced in Table 8.1, do not say much about success, quality or processes as does, e.g., the success of actions in defense or offense, the tactical quality of actions or the dynamics of the playing process (O'Donoghue & Holmes, 2014; Hughes & Franks, 2015; Memmert et al., 2017; O'Donoghue, Holmes & Robinson, 2017; Perl & Memmert, 2017). One reason is that quality or success is based on a combination of several attributes. As in the example on mean income above, the number and mean length of passes do not help to measure the quality of a player unless they are combined with each other or with additional context information about the situation or the corresponding tactical groups. However, the problem with combining attributes of different performance components to a complex performance attribute vector is that such multidimensional KPIs are not easily comparable anymore. In this case, the above-mentioned neural networks with their clustering concept can be helpful, as may be explained by the following example from governmental sports funding.

Example: sports funding

Sport needs money. The money is used for paying coaches and organization, financing infrastructure and travel or reimbursement of athletes. Some of the money can be received from entrance fees, advertising or private sponsoring. Any remaining funding is habitually expected from government (e.g., in Germany) and, often, the government has a special budget for partly financing disciplines in sports—on one condition, however: They have to be successful.

Without discussing the old problem of the chicken and the egg, i.e., whether funding enables success or success attracts funding, the basic problem is that of defining a discipline KPI based on a set of success attributes. For instance, such success attributes can be the number of gold, silver and bronze medals, positions in championships and rankings, number of athletes, infrastructure and coaching quality, media presence and public resonance. And immediately the question arises whether and how a deficit in medals could be compensated for by the number of athletes or the quality of the infrastructure (O'Donoghue, 2014, 2015).

The normal way is to weight the attributes by their importance and calculate the discipline KPI as a weighted mean value. Obviously, as discussed above, this is not the best way. Additionally, in practice, it will be very difficult to explain why a discipline A with a KPI of 10.31 lacks funding while discipline B with a KPI of 10.32 is getting it.

Again, because they deal with complex patterns and avoid pseudo-objective precision, Fuzzy and neural network approaches seem to be better alternatives for making such calculations and decisions fair and transparent. In addition, more complex pattern information lends itself to neural network analysis rather than statistical analysis.

Concepts and methods of modelling

As discussed in Part II, modelling KPIs not only means counting situations or events but also ensuring that, for a given KPI, the KPI values of different objects are comparable and consistent under semantic interpretation. In the following, the method of Fuzzy modelling is briefly introduced below, and on p. 152 the basic concepts and methods of net-based pattern recognition are discussed based on the example of sport funding. On p. 154, the example of soccer is dealt with. Finally, on p. 159, the problem of evaluating KPIs is outlined.

Fuzzy modelling

As outlined in Perl (2015):

> Fuzzy modelling first of all has to do with understanding and describing a problem rather than with finding a solution. As one example from games, the attempt of finding a useful optimal strategy for a specific situation might fail because of unsuitable problem descriptions and/or unreasonable demands on the results – e.g. too high numerical accuracy: Although from the technical point of view it is possible to appoint a player's position on the ground with a deviation of only some centimeters, it makes no sense to use such over-precise numeric information in order to describe 'optimal' behavior in a way like 'take the first serve on position [-0.25; +1.57]'.
>
> (Perl, 2015, p. 138)

Instead of such apparent precision, Fuzzy modelling tries to use imprecise terms in the same way that human communication does which, by the way, is the very reason for understanding each other. For instance, in the case above, the tennis coach could suggest taking the serve about a foot behind the base line and about a long step to the right of the center line. In terms of Fuzzy modelling, "a foot behind the base line" could read "a range from +0.10 to -0.60 with a degree of membership varying from 0.1 for '+0.10' up to 0.7 for '-0.30' and down to 0.1 for '-0.60'."

More information about the history of the Fuzzy approach and application in sports is given in Kosko (1992), Zadeh (1965), Zimmermann (1991) and Perl (2008).

Applied to the problem of measuring quality using KPIs, this approach has different consequences. In the case of passing players from p. 148, the typical situation is sketched in Figure 8.1. Three overlapping intervals represent quality classes of pass numbers and, accordingly, overlapping functions represent the degrees of membership of pass numbers to these quality classes. As has been outlined on p. 148, the player A with 19 passes and the player B with 20 passes with different degrees of membership are members of both the quality classes "good" and "excellent." One also can say A and B belong to the same quality cluster "good–excellent."

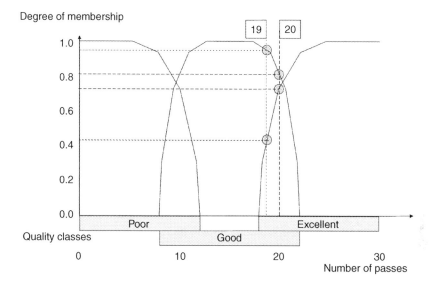

Figure 8.1 Fuzzy quality classes regarding the number of passes. The pass numbers 19 and 20 belong to the same classes "good" and "excellent" with different degrees of membership.

However, one important methodical fact that has to be kept in mind if working with the Fuzzy approach is that the decisive functions of degree of membership do not fall from heaven but have to be designed by experts. Therefore, in the end, this design determines membership of a certain quality cluster.

The situation becomes more difficult if there is not just one criterion like "number" but a more-dimensional vector (like number, mean length) that determines the quality ranking. The problem is that the quality classes of the components can be different. What happens if, e.g., player A has 19 passes with a mean length of 25 m and player B has 20 passes with a mean length of only 12 m? Based on the Fuzzy approach, one could say that, regarding the number of passes, players A and B are similar. Regarding the lengths of passes, player A definitely belongs to a better class than player B. Therefore, A is better than B.

But what about A = (8.25) and B = (20.13)?

The values of A and B are not in the same quality class, neither regarding the number of passes nor regarding the mean length of passes.

In such situations, the mean-value solution can very often be found where the mean values of the respective component values are compared and used for ranking. In the example, A and B have the same mean value 16.5, which means that their pass qualities are identical. Of course, such a mean value is a methodological nonsense, because the components can have quite different semantics and scales. But by reducing them to the mean value, they can compensate each other in an uncontrollable way.

This problem becomes substantial, e.g., in the case of sport funding from above where quite different criteria like medals, positions in championships and rankings, number of athletes, infrastructure and coaching quality, media presence and public resonance decide on the membership of funding classes.

One way to solve this problem is the method of clustering, i.e., grouping similar attribute vectors into clusters where, afterwards, all members of a cluster are treated equally. Therefore, the method of clustering can be understood as a kind of multidimensional Fuzzy-classification.

The approach sounds easier than it is. Take, e.g., the colors of the rainbow, where yellow is similar to orange, orange is similar to red, red is similar to violet, violet is similar to blue, blue is similar to green and, finally, green is similar to yellow. Obviously, similarity cannot be decided this static way but has to be developed dynamically based on the available set of objects and their multidimensional distribution. Self-organizing maps (SOM) have proved to be an excellent tool for doing this clustering.

Neural networks of type self-organizing maps

As outlined in Perl (2015):

> A self-organizing map consists of neurons, which are arranged in a matrix. [...] During the learning process, the input information triggers the particular target neuron the entry of which is closest to the input. The entries of the target neuron and, with decreasing weight at decreasing degree of neighbourhood, the entries of neighbouring neurons are moved directed to the new input. Due to those similarity-controlled neuron movements, eventually the distribution of the neuron entries is topologically similar to the distribution of the input data.
>
> After the learning phase, clusters of neurons represent classes of similar input data and so can classify types of data. This way, the high variety of data can be reduced to a comparably small number of characteristic types – e.g., the enormous variety of attacking processes in soccer could be reduced to about 20 characteristic patterns.
>
> In general, the approach of self-organizing maps is useful if the aim is typing and clustering of complex data spaces as for instance in motions or in games: [...]
>
> (Perl, 2015, p. 143)

"Or, for example, in sport funding," could be added. In the presented example, the qualitative properties of sports disciplines are encoded by 17 attributes, measuring past success like medals or positions in championships, present infrastructure and coaching conditions and expectations in future developments. After having been trained with those 17-dimensional attribute vectors, the neural network shows a result as presented in Figure 8.2.

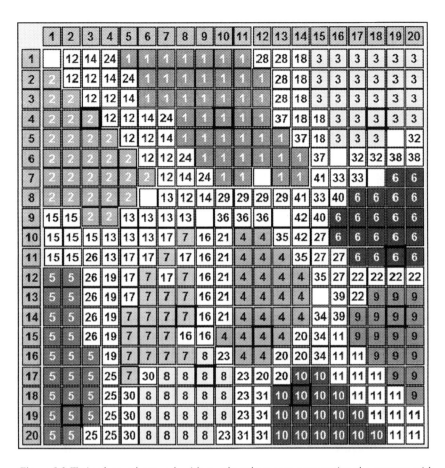

Figure 8.2 Trained neural network with numbered squares representing the neurons with the cluster numbers they belong to. The gray-shaded neurons represent the 10 largest clusters. The prototypes or most representative neurons of those clusters are highlighted by black marks.

The used network consists of 20×20 neurons, arranged in a square. The neurons are represented by small squares the numbers of which denote the clusters they belong to. As can be seen in this particular case, the network recognized 42 clusters, each representing a particular type of attribute pattern. Most of those clusters are very small, representing more or less unimportant patterns. The ten greatest clusters, representing the most important patterns, are highlighted by colors. In addition, each colored cluster shows one particularly marked neuron that is the most representative neuron or prototype of the cluster.

In order to classify the attribute vector of a discipline, the vector is fed to the net, and the net identifies the most similar prototype by comparing the distances

Discipline	1								Nr.	9		i	19		j	14	
Attribute	1	2	3	4	5	6	7	8	9	10	11	12	13	14	15	16	17
Prototype	6,7	9,5	3,4	9,4	8,8	9,5	7,8	5,8	0,5	9,5	6,3	4,4	8,2	8,6	8,0	7,7	8,5
Discipline	6,7	10,0	3,6	10,0	9,4	10,0	7,8	6,0	0,0	10,0	6,0	4,0	8,0	8,8	7,6	7,6	8,7
Difference	0,0	-0,5	-0,2	-0,6	-0,6	-0,5	0,0	-0,2	0,5	-0,5	0,3	0,4	0,2	-0,2	0,4	0,1	-0,2

Figure 8.3 Attribute table comparing the attributes of the selected discipline 1 and the correlated most similar prototype 9, together with the respective attribute differences. The numbers i and j denote the coordinates of the prototype on the neuron matrix (see Figure 8.2).

between the discipline vector and the prototype entries, as shown in Figure 8.3 for discipline 1.

Figure 8.3 shows a table with the attribute values of discipline 1 in the discipline line, the corresponding values of the identified best-fitting prototype 9 (with the neuron coordinates (19,14)) in the prototype line and the differences between prototype and discipline attributes in the difference line. Obviously, the correspondence between prototype and discipline is rather perfect.

So far, the process is a pure syntactic one, based on similarity and automatic clustering. For the decision about the amount of funding, the clusters or prototypes need semantic interpretation, e.g., by means of predicates like "poor," "good," "excellent" or, for more differentiation, by means of weight values. In both cases, these semantic values can be set based on expert assessment or calculated by means of algorithms developed on the basis of expert knowledge, or as a combination of both.

The main advantage of this concept is that not one single discipline has to be assessed—which could generate a lot of trouble between disciplines and experts— but the prototypical patterns, i.e., the number of which is small and allows for an intensive and comparing discussion independent of particular discipline interests.

To sum up: The conceptual idea is to characterize the profile of the discipline by a vector of first-level KPIs in a first step, followed by mapping this KPI-vector to a second-level KPI the shapes of which are represented by the patterns or prototypes (this approach has been developed by one of the authors (Perl) and will be used as PotAS software in German Sports Funding).

More information about the history of neural networks and their application in sports is given in Hopfield (1982), Kohonen (1995), Leser (2006), Memmert & Perl (2006, 2009a, 2009b), Perl and Dauscher (2006), Pfeiffer and Perl (2006) and Perl and Memmert (2011, 2012).

Example: soccer

In soccer, several KPIs are used to measure performance of players or the team. They either just count frequencies or calculate mean values or statistical

Table 8.2 Some examples of player- and team-oriented KPIs, categorized by mean value, frequency and distribution.

Examples	Player	Team
Mean values	running speed path length outplayed players per pass	path length outplayed players per pass duration of ball recovery ball possession rate
Frequency	number of: ball losses ball recoveries passes initialized attacks	number of: ball losses ball-recovery processes passes attacking processes
Distribution	locations of ball recoveries areas of ball control locations of ball losses passes from–to areas and rates of space control	locations of ball recoveries areas of ball control locations of ball losses passes from–to areas and rates of space control

distributions. Some KPIs of increasing complexity are listed in Table 8.2 (also compare Table 8.1).

More informative than indicators that just count single events, like passes or situations such as ball recovery, are those indicators that are geared toward the playing process and are therefore able to model and to reflect the dynamics of the play.

One example of a process-oriented KPI, which also uses neural networks for recognizing patterns, deals with formations and is presented in detail and discussed in the chapter on soccer. A formation is a geometric pattern of the relative positions of players of a tactical group like offense or defense. The relative positions of the players and therefore the pattern of the positions change depending on the acting intention and the behavior of the opponent group. The numbers of such interactions—which type of formation against which opponent type of formation—can be put into a matrix and so provide information about tactical behavior and chances (also see Part IV).

Another example of a process-oriented KPI, which is also presented in the chapter on soccer, measures the quality of passes by embedding the pass-event into the pressure situation as a specific process context. Three aspects or components are used to model the pass quality, namely: (1) pressure at the moment of passing the ball, i.e., the distance to the closest opposing player; (2) pressure at the moment of receiving the pass—again, the distance to the closest opposing player; and (3) the number of outplayed opposing players. The idea of modelling is that the quality of a pass (or better: the ability to make a pass) increases with (1) and (3) but decreases with (2), because high pressure at the moment of receiving the ball significantly reduces the chance of continuing the process.

The last example of a process-oriented KPI is an approach where two simple situational KPIs are combined to a KPI that measures the attacking efficiency of a team in a rather simple way.

One KPI is control of the ball that normally has to be recorded manually online from the play or offline from a video of the play. Another KPI is the control of space that is determined by means of Voronoi cells that, at any point in time, is the set of all points of the pitch a player can reach faster than any other player. As outlined in more detail in the chapter on soccer (p. 83; also, see Fonseca, Milho, Travassos & Araújo, 2012; Taki & Hasegawa, 2000), Voronoi cells help to measure the player-specific percentage rates of space control. In the context of attacking, in soccer, the 16 m or penalty area and the 30 m area in front of the opponent's goal are the areas of most interest (see Memmert & Raabe, 2017; Memmert et al., 2016a, 2016b).

The goal of this approach is to demonstrate how an appropriate combination of ball and space-control values leads to an indicator that provides quantitative information, like frequencies and distributions of events, without losing the qualitative information of the dynamic playing process. In what follows, the term "success," as it is usually used in an economics context, is implemented into a model of offensive success in soccer. The examples presented in figures 8.4 to 8.7 are calculated with the soccer analysis software SOCCER©J.Perl, which contains numerous components for a wide range of statistical and dynamic analyses of processes in soccer (see Perl & Memmert, 2011; Perl, Grunz & Memmert, 2013).

In Figure 8.4, the defense area of the defending team A against the attacking team B is presented. The Voronoi cells of the players of team A are light gray, those of team B are dark gray. The cells correspond to players whose numbers are noted in the small boxes.

A time-depending graphical representation of both the ball-control events and the space-control rates of team A and team B is given in Figure 8.5. It can be seen easily how the maxima of space control (SC) during attacks change between team A and team B, and how ball control (BC)—often, but not always—corresponds to space control. While team B combines high SC rates with BC in most of the attacking phases, team A does not. Although they control the space in the attacking areas impressively, it is useless because they do not control the ball, i.e., most of their attacks are not efficient if a coincidence of BC and SC seems to be necessary to make an attack dangerous.

Attacking efficiency over a time interval I may therefore be defined as correlation between ball-control events $BC(t)$ and space-control rates $SC(t)$ for $t \in I$, which leads to

$$\text{Efficiency}(I) = \text{Corr}(BC(t), SC(t)), t \in I.$$

A more detailed analysis and deduction is given in Perl and Memmert (2017).

A basic formula from economics containing the term "success" leads from efficiency to success:

$$\text{Efficiency}(I) = \text{Success}(I)/\text{Effort}(I),$$

where the term "effort" in the case of attacking soccer obviously can be defined as the number of all activities that together generate dangerous situations in the

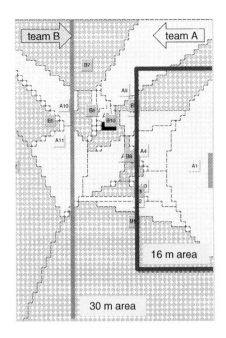

Figure 8.4 Voronoi cells of the players of the attacking team B (dark gray) and the defending team A (light gray) with the numbers of the respective players. The black square behind player B10 symbolizes the location of the ball.

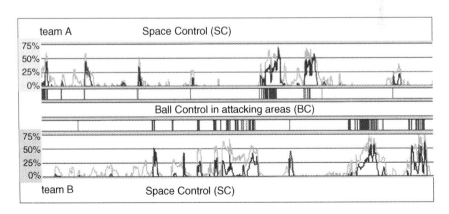

Figure 8.5 Time-diagram of a play (part of one half) showing profiles of the space-control rates in the opponent's 30 m (gray) and 16 m (black) areas as well as the ball-control events (vertical gray lines above (team A) and below (team B) the text "Ball control in attacking areas (BC)").

opponent's critical area over a selected time interval. This equation can easily be transformed to a model defining success depending on efficiency and effort:

Success(I) = Efficiency(I) × Effort(I).

With "effort," the sum over all activities regarding BC or SC, i.e.,

Effort(I) = sum(BC (t) ⊕ SC (t)), t∈I,

the attacking success over the time interval I can be calculated by

Success(I) = Corr(BC(t), SC(t)) × sum(BC (t) ⊕ SC (t)), t∈I.

(In order to avoid double-counting active t-points, instead of "+" the operator "⊕" is used with a meaning similar to the logic "or," i.e., "1+1=1.")

Figure 8.6 shows an example containing the profiles of the SC rates of the 16 m areas and the 30 m areas as well as the BC events. Over the marked time interval of 300 s, team A has an acceptable success value of 0.43 (maximum value = 1), while the success value of team B for the same time interval is close to 0, which is not surprising because, during the attacking phase of team A, team B was busy defending. The black profiles on top and on bottom of Figure 8.6 show the courses of the 300 s success values of team A and team B, respectively. Obviously, team A is much more successful in attacking than team B.

The advantage of that approach is that it calculates the KPI "success" depending on definable intervals, therefore allowing a process-oriented dynamic evaluation.

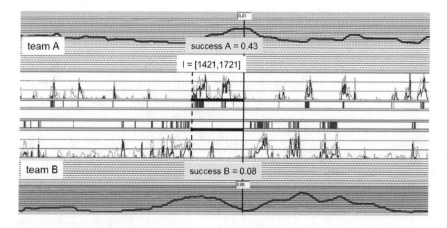

Figure 8.6 Time-diagram showing the success values of team A and team B at time 1,721 s = 28:41, calculated for the interval of the respective last 300 s (marked by dark gray bars). The black profiles at the top and bottom show the continuous course of those success values of teams A and B, respectively.

Of course, such complex indicators like formation interaction or attacking success are not as easy to handle as numbers of passes or percentages of BC but need additional analyses and interpretation. In the end, however, such dynamic KPIs provide much more information about the play and the players than simple numbers.

Problem validation

Finally, after modelling, it has to be verified that the KPI is able to deliver useful information about the modeled system. This is normally done by means of correlation analysis, which, however, has to be conducted carefully. The reason is that correlation analysis measures the quantity of similarity between the correspondent values of value sequences; it does not give logic information like "A implies B" if A correlates with B. That means that a great number of passes in a soccer game may highly correspond with a good result, but without guaranteeing that an increasing number of passes would improve the results (also see the example of tiki-taka in Chapter 4).

Methodological experience

As pointed out earlier, the most challenging task is to characterize a system with complex dynamics by just one or only a handful of key parameters.

But even if an appropriate model is available, the usefulness of which can be proved mathematically or statistically, the problem of getting useful information is not solved. Taking the example of a soccer play, different levels of data and information have to be distinguished regarding their availability, their types, the usefulness of the contained information and the ways of developing the contained information from the recorded data.

The best-developed technique of recording data from a soccer play is recording the position data of the players, where several technical solutions allow to automatically record data with satisfying accuracy. The positions of the ball are not quite so easy and often have to be added by manual recording. The recorded raw data can be used for developing information ranging from running performance to the behavior of tactical groups. In particular, the example of "pass under pressure" (see p. 155 and the chapter on soccer) demonstrates a direct way from recorded data to useful information that helps to define a model and a KPI of passing success. The reason is that, in this example, the change from available data to useful information is comparably easy and does not need semantic interpretation. The only question is which player is making a pass in which second. The necessary information can be found by manual data recording or by means of appropriate analysis software. Significantly more difficult is the example of "offensive success" from p. 158. In this case, "space control" can easily be calculated using Voronoi cells. But to calculate "ball control" is not quite so easy, being based on some semantic interpretation of what "control" actually means. At least, the example showed a way of defining "success" in a formal way, using reliable and accepted models.

Leaving the level of easily available information with no or little need of semantic interpretation, the task of defining appropriate KPIs immediately becomes much more difficult. First of all, the term "success" is a main problem. Of course, the main measure of success is winning the game, achieved by one team scoring more goals than the opposition. That means the main success of winning the game has two sub-successes, namely, to score as many goals as possible and to prevent (as much as possible) the opposition scoring goals. This approach can be continued in a stepwise refinement from the success of shots on goal down to the success of attacks, down to the success of contained passes and, finally, to the success of winning and keeping possession of the ball—all that embedded in the success of the contextual tactical patterns. The success of a component on a higher level always depends on the interaction of success on lower levels.

For example, assume a team of players who are technically perfect but unable to read situations and processes. Again and again, they win the ball and make wonderful passes but fail to recognize the right moment to mount an attack and, as a consequence, fail to develop any threat on goal. In turn, an experienced chess player might be able to read the play and to make the best tactical decisions but fails to get their tactics to work against experienced and resolute opponents.

The consequence is that it makes no sense to measure isolated success values if they are not imbedded in or based on processes. KPIs dealing with such isolated success parameters are not completely useless. If, e.g., most of the players are too slow or tired after 60 minutes, or if most of the passes are too short or do not reach receiving players, then there are obviously indicators found saying that improvement is needed for the chance of winning a game. But even with improving such isolated success parameters there is no guarantee of more success unless cooperation among the players is also improved, i.e., such KPIs are not necessarily helpful in predicting what has to be improved in order to increase the chances of winning the game.

As an example, take Ronaldinho, an attacking player for Futbol Club Barcelona around 2005. He moved slowly and he did not even move much. But, at the right moment, he was in the right place, fast in his action and shot on goal. What he needed were teammates who shared his understanding of the game for perfect passes and timing. Today, Ronaldo from Real Madrid plays a similar role, although his speed and engagement in the match is quite different.

Finally, there is one important problem in getting and measuring success. This problem arises from the activities of the opposing team. A team cannot be successful with passes if the opposition prevents them. This problem has been addressed intensively since the 17th century (not in the field of soccer, however) and led to the mathematical discipline of game theory. Besides a lot of statistical results concerning gambling, one remarkable theory is that of two-person zero-sum games that—a bit simplified—says that the success of one player is always limited by the success of the other player.

Transferred to soccer, this means that interaction with the opposing team plays an important role in measuring success of one's own team.

One example from soccer, which is dealt with on p. 155, is the coincidence matrix of formations that counts the number of coincidences of each pair of offensive formations of one team against defensive formations of the other team. If these coincidences can be combined with values of success, e.g., numbers of BC or percentage rates of SC, the matrix becomes a success matrix providing information about which tactical steps can improve success in which situation (see the chapter on soccer).

Another way of getting more information about the dynamic processes of a play is to work with time-depending profiles presenting activities of the one team in the context of the activities of the other team, as with BC and SC on p. 158. If those profiles represent parts of the dynamics of the play, they also reflect the interaction between the teams and can therefore make clear, e.g., when and how activities of the one team influenced the activities of the other team. Projected against this background, even simple KPIs like ball possession make sense because they are imbedded in the context of the interactive playing process.

Moreover, as demonstrated in Figure 8.7, such profiles can be combined interactively with animations or videos of the game in order to show the playing situation if at one point in time a striking feature is detected.

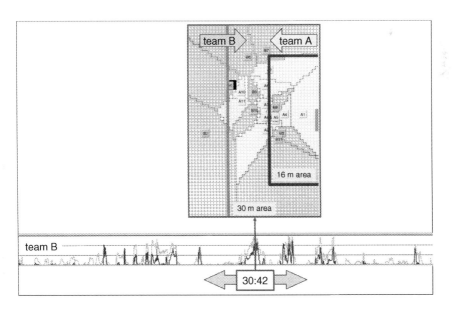

Figure 8.7 Interactive coupling of the time-diagram and animation or video components: Scrolling to the point of interest shows the corresponding playing situation, e.g., the distribution of space control.

Results

On pages 150–159 and 159–161, modelling aspects and usefulness of KPIs were discussed. It turned out that it is rather difficult to assess KPIs under aspects of modelling and scientific analyses. The situation becomes even worse if taking into account that KPIs are not primarily designed for scientific purposes but for practitioners with well-based background knowledge. If, for instance, a soccer coach reads the play of his team in complex patterns and evaluates it against the background of his experience, then the simple number of passes can possibly provide valuable information although it does not reflect the quality of processes. But it can give the coach insights to player constellations, tactical variants or relevance depending on the opposing teams.

By the way, if coaches need only simple KPIs because they are recording and interpreting complex patterns in a play, then this ability of pattern analysis is scientifically thrilling and should be subject to research.

Moreover, players' lack of ability in running or passing can be reasons for changing the training. Although those abilities alone are not sufficient for improving the quality of the play, as discussed in Methodological experience, such basic abilities are of course necessary to play on a not-too-low level. In modern soccer in particular, speed and conditioning are highly important in order to get fast attacks to work and exert pressure early on.

Experience gained from over 30 years of modelling in sports shows that the communication between coach and scientist can be extremely difficult; not only is it necessary to speak the language of the other but it is also necessary to understand each other's points of view and experience. Because of this problem, over the last few years, a new position in the coach's crew has developed: the IT assistant, whose task is to transfer knowledge, demands and results between the coach and IT-based analyses. It seems to be a good way of improving information by sharing it.

Example

One well-known example of a simple KPI, which has been influencing technical and tactical play in soccer over several years, is "ball possession," also known under the term "tiki-taka," which describes the endless sequences of passes in the midfield in order to keep the ball under control.

The idea is that if one team controls the ball, the opposing team cannot score goals. But the justified question is how the team controlling the ball can score goals if the ball simply circulates around the midfield. The answer is: wait for a good chance. Indeed, at the beginning of the tiki-taka age, teams that were perfect in passing and controlling the ball were often winning their matches. However, when the effect of surprise disappeared after a while, it turned out that teams with good counter-attacks were able to defeat a tiki-taka strategy, and that even with high goal differences.

From a modelling point of view, the question is, therefore: What are the reasons behind success and failure of ball-possession-based tactics?

Only some rather simple KPIs are needed:

(1) Let BPR (Ball Possession Rate) be the percentage of time a team controls the ball. As discussed above, passing in the midfield is not really dangerous for the opposing team. Therefore,
(2) Let ER (Exploiting Rate) be the percentage of ball possession time when a team generates dangerous situations. Of course, not all of the generated dangerous situations are successful, where "successful," e.g., could mean a shot on goal or a standard situation in the attacking area. Therefore,
(3) Let SV (Success Value) be the percentage of being successful, i.e., $SV = 100 \times (ER/100) \times (BPR/100)$.

Comparing team A with the success value SVA and Team B with SVB calculates the advantage of A against B as

$$AdvA = SVA - SVB.$$

Some examples of values of AdvA depending on different constellations are presented in Figure 8.8.

It turns out that the tiki-taka team A needs an extremely high rate of ball possession (75%) to have a chance against a team B, which uses its low rate of ball possession (25%) for a great rate of dangerous counter-attacks (20%).

The diagram of the advantage AdvA of team A depending on its ball possession rate BPRA is shown in Figure 8.9.

To complete the example, in the following, the idea is briefly presented how the value BPRA of tiki-taka team A can be calculated, which is particularly necessary to have a chance to win against counter-attack team B.

Condition $SVA \geq SVB + S$ where S is the safety-value 2.5 from Figure 8.9 Implies

$$(ERA \times BPRA)/100 \geq (ERB \times BPRB)/100 + S,$$

BPR A = 55% and ER A = 10% ⇒ SV A = 5.5%		
BPR B = 45% and ER B = 20% ⇒ SV B = 9.0%	⇒	SV A - SV B = -3.5
BPR A = 67% and ER A = 10% ⇒ SV A = 6.7%		
BPR B = 33% and ER B = 20% ⇒ SV B = 6.6%	⇒	SV A - SV B = 0.1
BPR A = 75% and ER A = 10% ⇒ SV A = 7.5%		
BPR B = 25% and ER B = 20% ⇒ SV B = 5.0%	⇒	SV A - SV B = 2.5

Figure 8.8 Examples of success values of team A and team B (SVA, SVB) and the advantages of team A (= SVA–SVB) depending on respective exploiting rates (ERA, ERB).

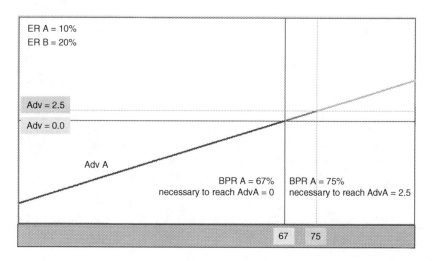

Figure 8.9 Diagram showing the values of AdvA = SVA − SVB depending on ERA = 10% and ERB = 20%. The result is that team A needs 67% of ball possession to get a balanced situation and 75% to get an advantage of 2.5.

and some simple transformations lead to

$$BPRA \geq (100 - BPRA) \times (ERB/ERA) + (100 \times S)/ERA.$$

Here, the fact is used that BPRA + BPRB = 100, i.e., the percentage parts of time have to add to 100.

From this, it finally follows what the minimal value of BPRA has to be:

$$BPRA \geq 100 \times (ERB + S)/(ERA + ERB).$$

To get a final idea of what that means, take, e.g., ERA = 5%, ERB = 20% and S = 0. The result is that BPRA must be at least 80% to have a chance of not losing. And taking S = 2.5 to maintain a chance of winning increases the necessary value of BPRA to 90%. This means that on a bad day (i.e. low value of BPR), team A not only has no chance of winning but, moreover, has a good chance of losing with a significant negative goal difference.

Exercise

The exercise deals with the tiki-taka example from Chapter 4. The final result developed above calculates the minimum value of the BPR of team A that maintains a chance of winning the game.

In order to get familiar with the structure of the formula, the first task is to transform the result in such a way that the minimum value for team B can be

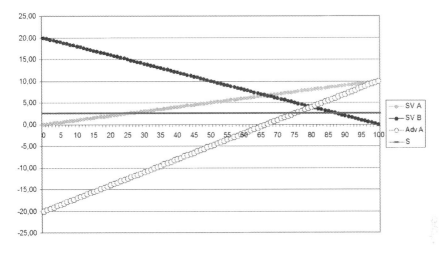

Figure 8.10 Excel diagram showing SVA (gray), SVB (black) and the resulting values of advantage for A (white). The black horizontal line shows the advantage value S = 2.5.

calculated respectively. Also, it would be interesting how the values of ERA and ERB depend on BPRA, BPRB and S.

The second task is to develop an Excel application similar to the one in Figure 8.10, where the values of the parameters ERA, ERB and S can be entered, affecting the behavior of the resulting curves.

The third and final task is to discuss the results in the context of what is possible and happening under realistic circumstances.

References

Fonseca, S., Milho, J., Travassos, B. & Araújo, D. (2012). Spatial dynamics of team sports exposed by Voronoi diagrams. *Human Movement Science, 31*, 1652–1659.

Hopfield, J. (1982). Neural networks and physical systems with emergent collective computational abilities. *Proceedings of the National Academy of Sciences, 79*, 2554–2558.

Hughes, M. & Franks, I. (Eds.). (2015). *Essentials of performance analysis in sport.* Abingdon: Routledge.

Kohonen T. (1995). *Self-organizing maps.* Berlin; Heidelberg; New York: Springer.

Kosko, B. (1992). *Neural networks and fuzzy systems.* Englewood Cliffs: Prentice-Hall.

Leser, R. (2006). Prozessanalyse im fußball mittels neuronaler netze [Process analysis in football by means of neural networks]. *Human Performance and Sport, 2*, 199–202.

Memmert, D. & Perl, J. (2006). Analysis of game creativity development by means of continuously learning neural networks. In E. Moritz & S. Haake (Eds.), *The Engineering of Sport 6, 3,* 261–266. New York: Springer.

Memmert, D. & Perl, J. (2009a). Analysis and simulation of creativity learning by means of artificial neural networks. *Human Movement Science, 28,* 263–282.

Memmert, D. & Perl, J. (2009b). Game creativity analysis by means of neural networks. *Journal of Sport Science, 27,* 139–149.

Memmert, D., Lemmink, K. & Sampaio, J. (2017). Current approaches to tactical performance analyses in soccer using position data. *Sports Medicine, 47*, 1.

Memmert, D., Raabe, D., Knyazev, A., Franzen, A., Zekas, L., Rein, ... Weber, H. (2016a). Big data im profi-fußball. Analyse von positionsdaten der fußball-Bundesliga mit neuen innovativen Key Performance Indikatoren [Big data in professional football. Analysis of position data from the German Bundesliga with new innovative Key Performance Indicators]. *Leistungssport [High Performance Sport], 5*, 1–13.

Memmert, D., Raabe, D., Knyazev, A., Franzen, A., Zekas, L., Rein, R., ... Weber, H. (2016b). Innovative leistungsindikatoren im profifußball auf basis von positionsdaten [Innovative performance indicators in professional football based on position data]. *Impulse [impulse], 2*, 14–21.

Memmert, D. & Raabe, D. (2017, forthcoming). *Revolution in professional soccer:* With Big Data *towards match a*nalysis 4.0. Berlin: Springer-Verlag.

O'Donoghue, P. (2014). *An introduction to performance analysis of sport*. Abingdon: Routledge.

O'Donoghue, P. & Holmes, L. (2014). *Data analysis in sport*. Abingdon: Routledge.

O'Donoghue, P. Holmes, L. & Robinson, G. (2017). *Doing a research project in sport performance analysis*. Abingdon: Routledge.

Perl, J. (2008). Modeling. In P. Dabnichki & A. Baca (Eds.), *Computers in sport*. Southampton; Boston: Wit Press.

Perl, J. (2015). Modeling and simulation. In A. Baca (Ed.), *Computer science in sport* . London; New York: Routledge.

Perl, J. & Dauscher, P. (2006). Dynamic pattern recognition in sport by means of artificial neural networks. In R. Begg & M. Palaniswami (Eds.), *Computational intelligence for movement science* (pp. 299–318). Hershey; London; Melbourne; Singapore: Idea Group Publishing.

Perl, J. & Memmert, D. (2011). Net-based game analysis by means of the software tool SOCCER. *International Journal of Computer Science in Sport, 10*, 77–84.

Perl, J. & Memmert, D. (2012). Network approaches in complex environments. *Human Movement Science, 31*(2), 267–270.

Perl, J. & Memmert, D. (2017). A pilot study on offensive success in soccer based on space and ball control: Key Performance Indicators and key to understand game dynamics. *International Journal of Computer Science in Sport, 10*, 77–84.

Perl, J., Grunz, A. & Memmert, D. (2013). Tactics in soccer: An advanced approach. *International Journal of Computer Science in Sport, 12*, 33–44.

Pfeiffer, M. & Perl, J. (2006). Analysis of tactical structures in team handball by means of artificial neural networks. *International Journal of Computer Science in Sport, 5*(1), 4–14.

Rein, R. & Memmert, D. (2016). Big Data and tactical analysis in elite soccer: Future challenges and opportunities for sports science. *SpringerPlus, 5*(1), 1410.

Taki, T. & Hasegawa, J. (2000). Visualization of dominant region in team games and its application to teamwork analysis. In *Computer Graphics International, 2000*. Proceedings (pp. 227–235). IEEE.

Zadeh, L. (1965). Fuzzy sets. *Information and Control, 8*, 338–353.

Zimmermann, H-J. (1991). *Fuzzy set theory – and its applications* (2nd edn.). Boston, MA: Kluwer Academic Publishers.

Part IV
Physiological conditions of being successful

9 Marathon

Stefan Endler

Introduction

According to legend, a Greek warrior ran about 40 km to proclaim the victory of a battle and died after his proclamation. This legend was taken up in 1896 at the first Olympic Games of the modern era. A long-distance run of about 40 km was created and called a marathon. Nowadays, 42.195 km is the official distance of a marathon. This distance was set at the Olympic Games in London in 1908 because of the chosen start point at Windsor Castle and the finish line at the Olympic stadium. For many decades, a marathon was a distance more for professional athletes and ambitious amateurs. However, in the last 3 decades, more and more people have tried to finish this distance, which is one of the most difficult challenges in sports altogether.

Especially for beginners, the main challenge is the lengthy preparation. Muscles and energy supply must be adapted to a long-distance run, meaning constantly running for a very long period without rest. Therefore, many weeks of training is required. High perseverance and ambition are needed to execute such a program. However, even a well-executed training program is no guarantee for a successful marathon. During the competition, many external factors, e.g., spurring on of spectators or weather, influence the body. The biggest challenge is finding an optimal speed level; starting out too fast would lead to an early breakdown, also known as "hitting the wall," however starting out too slow will not exploit the athlete's full potential.

For an athlete, it is helpful to know an approximate finish time to avoid exhausting their energy reserves and face an early breakdown. Knowing an optimal marathon finish time can be used to prescribe an individual pacing strategy. Several different techniques can be used for that approximation. The easiest way is a competition run over a shorter distance before the marathon. The marathon finish time can then be extrapolated using the finish time of the shorter distance competition. Therefore, only time must be measured. Other techniques use load parameters, e.g., running speed in km/h or min/km and parameters measuring reaction of the body due to load, e.g., heart rate (HR), lactate or oxygen uptake. Some parameters can be measured continuously. These parameters can even be used for further optimization during a competition.

Others still require a higher effort due to blood extraction or cumbersome equipment. These parameters are restricted to use just before a competition and often only in a laboratory setting.

All techniques use models for their approximation of finish time. Some models use many data of a supposed basic population. Others try to individualize and, thus, represent typical physiological phenomena that can be adapted to individual athletes. These approaches have the benefit of explaining performances physiologically. Individualized to an athlete, models can be used to determine or simulate an optimal marathon finish time. This chapter will show several of these models and how they approximate the finish time of a marathon. A core theme will be the antagonistic model PerPot (Performance Potential metamodel), which was adapted to a running-specific model, PerPot-Run. It uses the parameters HR and speed, hence it is usable as an online optimization tool too.

Project/problem

Predicting the goal time of a marathon can help athletes to prescribe their pacing strategy. This should lead to exhaustion at the finish, and not earlier. Such a prediction is important, especially for beginners with less experience who tend to begin the marathon too fast, which typically results in hitting the wall after approximately 30 km.

Goal times can be predicted by two different types of models. The first uses big data. Based on the data of many athletes, (statistical) models can be built to approximate mean marathon goal times. The purpose of these models is the determination of correlations between initial data and marathon goal times. The first and most common model is the formula of Riegel (Riegel, 1981). All these models (e.g., Blythe & Király, 2016; Pradier et al., 2016) approximate the goal times based on an alleged basic population. Thus, the individuality of the athletes is not taken into consideration.

However, the second type of models uses individual data of athletes to approximate the goal time. These models describe physiological processes to conclude the individual performance. Models of physiological phenomena can be fed by individual data and are thereby adapted to the athlete. Simulations, based on the adapted models, provide individualized predictions. In contrast to the first type of models, these models describe physiological phenomena and are not only based on population means.

The second type can be used in two different ways. First, data is measured or collected ahead of the competition. Predictions can be made before the competition by analyzing the data retrospectively. The most common models are the lactate-based models. Using lactate values under increasing load shows an exponential behavior. The interesting point, where lactate building and elimination are in equilibrium, is determined by using different mathematical models (Svendahl & MacIntosh, 2003). Based on this threshold, goal times can be determined for half- and full marathons too. A similar method uses breathing gas to determine a ventilatory threshold (Péronnet et al., 1987).

Goal time can be predicted by this threshold as well. The disadvantage of these methods is the risk of changing conditions from the data collection day to the competition day. For example, changing weather conditions (Ely, Cheuvront, Roberts & Montain, 2007), especially temperatures and humidity, have a high influence on performance (El Helou et al., 2012). Furthermore, competition-specific factors are not considered in the determinations. One of these factors is the altitude profile of the running course. Typically, predicted times assume a flat track (although there are only a few competitions in long-distance running that are completely flat) and incline has an appreciable impact on performance (Ferley & Vukovich, 2015). Another factor during competitions is spectators. As Laborde, Dosseville and Allen (2016) stated, emotional intelligence has a high impact in sport and physical activity. Spectators, who cheer the athletes on, might have a positive (exaltation) or negative (anxiety) influence on the physiological response.

Second, measured data during a competition, e.g., HR and speed, can be used for online optimization. Therefore, the previously individualized and adapted model parameters would be adapted again using data on a daily basis. Thus, changing conditions do not matter in this scenario. Furthermore, physiological response on, e.g., incline and decline or cheering of spectators are considered implicitly. With the newly determined goal time, adaptations to current running speed can be communicated to the athlete during the competition. One model that can be used during a marathon is PerPot-Run. A special app version of this model makes usage of the actual HR and speed values and predicts actual goal times during the marathon.

Concepts and methods of Modelling

Generalized models

One of the first established models was that of Riegel (1981):

$$t_{long} = t_{short} \cdot \left(\frac{s_{long}}{s_{short}}\right)^{1.07}$$

Riegel deduced an exponential correlation between short- and long-distance goal times. The deduction was based on the world records of the 1970s and earlier. Given an actual goal time of a short distance (t_{short}), a goal time for a longer distance can be determined. $\left(\frac{s_{long}}{s_{short}}\right)^{1.07}$ can be interpreted as a constant, e.g., for a marathon $s_{long} = 42.195$ and for a 10 km test race $s_{short} = 10$, the constant factor is $\left(\frac{42.195}{10}\right)^{1.07} = 4.667$. Therefore, a timetable can be built with marathon prediction times based on 10 km times (see Table 9.1).

Table 9.1 Prediction of marathon goal times out of 10 km best time using Riegel's formula.

10 km time in Min.	Predicted marathon goal time
75	5:50:00
70	5:26:40
65	5:03:20
60	4:40:00
55	4:16:40
50	3:53:20
45	3:30:00
40	3:06:40
35	2:43:20
30	2:20:00

Table 9.2 Actual times for three athletes at three competitions.

	10 km	Half-marathon	Marathon
Athlete A	40:00	1:30:00	3:10:00
Athlete B	41:30	1:30:00	3:05:00
Athlete C	43:00	1:30:00	-

Note: single marathon value missing

Besides the disadvantage of models using data of population mean mentioned above, Riegel's model has a further disadvantage: The formula is based only on world records. Thus, it is not generalizable, especially for beginners.

This method was enhanced by Blythe and Király (2016). Their method is not only based on world records, but on a database of British runners' performances. The database[1] contains nearly 1.5 million performances in different competition lengths from 100 m up to marathon.

In the first step, they use the machine-learning model "Local Matrix Completion." Thus, missing performances of athletes can be approximated, because not every athlete competes in every distance. Let's assume three athletes performing in three different competitions, with one missing entry (see Table 9.2)

Mathematically, athlete C is a linear combination of athletes A and B. To get a prediction of athlete C's marathon, the equation

$$\det \begin{pmatrix} 2400 & 5400 & 11400 \\ 2490 & 5400 & 11100 \\ 2580 & 5400 & X \end{pmatrix} = 0$$

must be solved. In this example, all values are converted into seconds. As a result, x = 10,800 (seconds) meaning exactly 3 hours for the marathon for athlete C.

Based on all the data, Blythe and Király (2016) present a parsimonious low-rank model, which explains individual running times t in terms of distances s:

$$\log t = \lambda_1 f_1(s) + \lambda_2 f_2(s) + \cdots + \lambda_n f_n(s)$$

The components $f_1, f_2, \ldots, f_n$ are the same for every athlete and can be specialized for different groups depending on available data, e.g., all males between 20 and 30. $\lambda_1, \lambda_2, \ldots, \lambda_n$ are coefficients that take the athlete into consideration. Analyses based on the database show optimal forecasts for n = 3.

Another model for predicting marathon goal time was presented by Pradier et al. (2016). They made use of Bayesian nonparametrics (BNP) and the hierarchical Dirichlet process (HDP) to analyze running patterns out of split times during marathon competition. Subsequently, goal times can be predicted by given marathon split times during the first kilometers.

A BNP is a probabilistic model with the objective of modelling uncertainty. Data in real-world scenarios, e.g., marathon modelling, has uncertainty due to noisy measurements or because one only has access to information of a subset and not the information of the whole basic population. Distributions, e.g., the zero-mean Gaussian process, are building the basis (prior) for those models.

In the first step of their prediction, Pradier et al. (2016) build clusters of running patterns using their HDP approach. They used the time measurements of all marathon finishers of the New York City marathon from 2007 to 2011 for their analysis. Including different pacing strategies and running speeds, 46 clusters were built. Prediction of the marathon goal time using the split times can be done by: 1) Computing the posterior probability of the athlete running in the considered cluster; 2) projecting the last available accumulated time to a predicted goal time separately for each cluster; and 3) the overall prediction is a weighted median of all predictions of each cluster. The median made the predictions more robust against outliers compared to the mean, which was used first.

Individual models

The first individual method was the lactate diagnostic. The method is based on the metabolic parameter lactate that is measured during a graded incremental test. From the late 1970s until now, this method has become important for predicting and prescribing endurance performance (Faude, Kindermann & Meyer, 2009). The goal of the diagnostic is to identify the individual anaerobic threshold. Among others, marathon goal time is calculated based on that threshold.

Two models are applied in the method. First, a curve must be fitted. Lactate is measured between every step of a graded incremental test. This value combined with the actual load value, e.g., speed, are plotted as points. The values in between the measured data are approximated by a curve. Different mathematical functions can be used for this purpose. In practice, an exponential function

$$f(x) = a \cdot e^{bx} + c$$

and a polynomial function of third order

$$f(x) = ax^3 + bx^2 + cx + d$$

have been enforced. However, this has not yet been investigated and is one of the disadvantages of lactate diagnostics (Faude et al., 2009). *a*, *b*, *c* and *d* are the model parameters, which must be optimized. *c* of the exponential function and *d* of the polynomial function can be set as the measured basic lactate value in rest. All others must be optimized numerically so that the measured lactate values are fitted with a minimal error. The error values are calculated normally by the least squared error method:

$$\sum_i (f(x)_i - y_i)^2$$

with $f(x)_i$ as the simulated value and y_i as the measured lactate values for all measurements i. High deviations between measured and simulated values are penalized higher due to the quadratic term.

Afterwards, a second model is implemented to find the anaerobic threshold. As Faude et al. (2009) show, several different models exist for this determination. Some models, e.g., Keul et al. (1979), are based on data of studies and therefore only semi-individual. Others, e.g., the most common model in practice of Dickhuth et al. (1999), is based on theories about the lactate kinetic, especially behavior close to the threshold. To go into all models in detail would take us too far afield. Thus, only Keul et al.'s will be described representatively.

Figure 9.1 shows how the model operates. The light gray points (at the bottom) are the measured lactate values at each step and the light gray line is the approximated lactate curve. An exponential fitting was executed. The dark gray points and line (at the top) represent the measured and linear interpolated heart-rate values. Keul et al.'s model works as follows: A tangent is drawn at the lactate curve with exactly 51.34°. The point of contact is defined as anaerobic threshold. At that threshold, HR and speed at the anaerobic threshold can be read off. The angle was defined based on a study where correct thresholds were known (Keul et al., 1979).

The speed at the anaerobic threshold can be used to forecast the marathon goal time. Besides others, one big problem of the lactate diagnostic is the assignability from laboratory results to outdoor events, as e.g., Vergès, Flore and Favre-Juvin (2003) showed for cross-country skiing.

A similar method is the spiroergometry (Bunc, Heller, Leso, Šprynová & Zdanovicz, 1987). In contrast to lactate diagnostic, respired gas is analyzed instead of lactate. The methodology is nearly the same. First, a graded incremental test must be performed. During this test, the athlete wears a mask that can measure and analyze the different concentrations of the respired gas, e.g., O_2 concentration can be measured for breathing in and out. The difference of both O_2 concentrations multiplied with the number of breaths per minute (V_e) results in the oxygen uptake (VO_2).

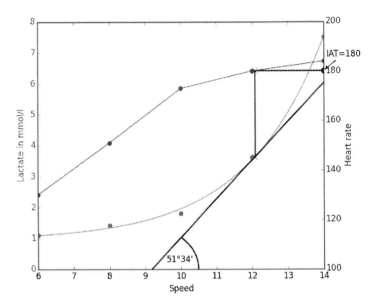

Figure 9.1 Keul's (1979) model.

Two thresholds can be determined: The aerobic threshold VT_1 and the anaerobic threshold VT_2, also referred to as the respiratory compensation point. VT_2 is defined as the point immediately below the VO_2 at which V_e continues to increase with time, rather than attain a steady state (Reybrouck, Ghesquiere, Weymans & Amery, 1986). Even today, determination of the threshold is computer assisted and not fully automated. Experts with lots of expertise have to visually inspect different plots of measured and determined data as time series because the threshold is not explicit. Algorithms are only able to support experts at the moment.

Another possible solution is the PerPot-Run model, which shapes the physiological phenomena of HR reaction under different conditions. Simulations can be used to determine marathon goal time. This basic model, as well as an extension using live data measured during the competition, are presented in the next section in detail.

Methodological experience (PerPot-Run)

Data acquisition

Models can only be verified and used in practice by simulations afterwards, if real data can be acquired. For the following approach of modelling and simulating marathon competitions, HR and speed are required. Both values are measurable in different ways; some of them will be described and discussed, including advantages and disadvantages.

HR and speed can be measured by different devices. Special HR monitors are constructed only for this purpose. Besides these two values, HR monitors measure other parameters, e.g., altitude or temperature. With modern devices like smartphones and smartwatches, data can be collected and displayed by apps. Mostly, the concrete measurement is not done by the devices themselves; special sensors measure the data and send it via Bluetooth, e.g., to the device.

Today, chest straps are used by default for measuring HR. Most chest straps allow for an EKG precise measurement as they measure the HR, beat by beat. Thus, not only HR is measured, but also HR variability, which can be used to improve training processes (Kaikkonen, Hynynen, Mann, Rusko & Nummela, 2010). However, one must familiarize oneself in wearing chest straps; in particular breathing of beginners is restricted by the strap. Furthermore, measurement of HR can be interrupted at times, e.g., the chest strap is not moist enough, and errors sometimes occur due to electric disruption, e.g., beside train lines.

One alternative provides optical sensors that are mostly integrated directly in the device. The flow of blood is measured using LED technology. However, this technology is still in its infancy, as Parak and Korhonen (2014) show for two different optical trackers. Here, measurement errors occur, e.g., because of the influence of daylight or movement of the device on the wrist.

The most common method of measuring actual speed during running is via global positioning system (GPS). The actual position of a person is tracked by the GPS using several satellites. However, for non-military usage, the precision is limited to ±10 m. Therefore, an accurate actual speed cannot be supported, as Figure 9.2 shows. Varley, Fairweather and Aughey's study (2012) shows coefficients of variance up to 11.3% even in straight-line running. By means of filters and smoothing algorithms, however, speed values might be improved afterwards. Another disadvantage is the high energy cost of GPS devices that must be recharged frequently.

A more precise method of measuring actual speed is a running sensor. Such a sensor is placed on the shoe and measures 3D accelerations and angles. The sensors can easily be adapted for use with different shoes. Furthermore, the sensors are very light and do not disturb the running itself. As specified by the most common manufacturers of endurance-sports monitors, the precision of these sensors is about ±3% after individual calibration. One disadvantage is their additional cost, since these sensors are not included in delivery of the HR monitors by default.

Model (PerPot-Run)

Modelling the HR process on the speed during a marathon can be divided into three phases:

1. Before the competition starts, HR begins at an individual level. Based on this starting level, HR grows delayed with an exponential attenuated reaction on a new speed level. This behavior shows up at every changing load,

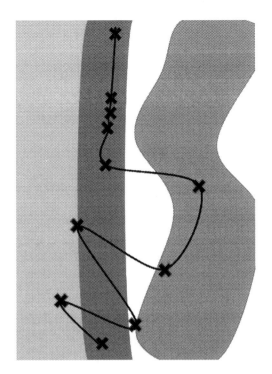

Figure 9.2 Schematic diagram of speed determination using GPS with ± 10 m variance.

mostly in changing speed conditions, but also in changing decline or incline conditions.
2. During the marathon, HR increases over the whole competition even at a constant speed level. This phenomenon is well known as cardiovascular drift (Heaps, Gonzalez-Alonso & Coyle, 1994).
3. The last phase is well known as "hitting the wall" and should be avoided absolutely. Hitting the wall is defined as the moment where glycogen supplies have been exhausted and energy must be converted predominantly from fat (Stevinson & Biddle, 1998).

The first and third phases are modeled by the PerPot (Perl, 2004). The metamodel was initially built to model training processes in an abstract manner, meaning the reaction of performance to training load. Besides the exponential delayed behavior and the overloading mechanism, PerPot also models physiological phenomena, e.g., the supercompensation effect. The metamodel was adapted to running (Endler, 2013). Phenomena that do not occur in HR reaction to speed, e.g., supercompensation, were removed. Additional phenomena were added, e.g., the second phase of continuous increasing of HR described above.

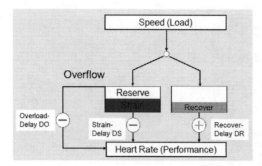

Figure 9.3 Structure of the performance potential (PerPot) model.

In general, PerPot is a deterministic and discrete computer science model conceived to represent interdependencies between load input and performance output. We use speed as load input and HR as performance output. The effect of running speed on HR results from two potentials. The recovery potential (RP) increments HR, whereas the strain potential (SP) decrements HR. Increment and decrement occur with different time delays, i.e., the recovery delay (DR) and the strain delay (DS). This basic model is able to show the behavior of the first phase described above.

Phase three, the overloading mechanism, is modeled by limiting SP. If running speed is getting too high, SP reaches its maximum and a further and shorter delayed negative effect to HR is provoked. This mechanism avoids an endless increase of HR (see Figure 9.3).

The difference equations of the interim values RP and SP and the output value HR are as follows:

$$RP\,[t + \Delta t] = RP[t] + \Delta t \cdot (\text{Speed}[t] - RR\,[t, t + \Delta t]) \tag{1}$$

$$SP\,[t + \Delta t] = SP[t] + \Delta t \cdot (\text{Speed}[t] - SR\,[t, t + \Delta t] - OR\,[t, t + \Delta t]) \tag{2}$$

$$HR\,[t + \Delta t] = HR[t] + \Delta t \cdot (RR\,[t, t + \Delta t] - SR\,[t, t + \Delta t] - OR\,[t, t + \Delta t]) \tag{3}$$

The three rates, i.e., strain rate (SR), recovery rate (RR) and overflow rate (OR), are determined by the following equations:

$$RR\,[t, t + \Delta t] = \min(1, RP[t], 1-PP[t])/DR \tag{4}$$

$$SR\,[t, t + \Delta t] = \min(1, SP[t], PP[t])/DS \tag{5}$$

$$OR\,[t, t + \Delta t] = \max(0, SP[t]-1)/DO \tag{6}$$

Phase two, the continuous increasing HR over time, is seen in linear increments of DS. Furthermore, as one can see, the rates could also result from performance potential (PP). This possible back-coupling causes a further phenomenon: An S-shaped HR response, given an incremental speed process (Brook & Hamley, 1972).

Individual calibration

At first, individual calibration is needed, since variation of running speeds causes different delayed HR reaction in individuals. Therefore, a graded incremental test must be performed to determine the individual delayed changes of HR related to changing running speed. Typically, the test starts with an initial workload of 6–8 km/h for 3 minutes and incremental increases of 1–2 km/h every 3 minutes until subjective exhaustion is reached with a minimum of five steps. After the last incremental step, another step of 3 minutes with the previous initial workload of 6–8 km/h takes place.

Running speed and HR process of the graded incremental test can now be used to adapt the model to the athlete. Therefore, internal model parameters, e.g., delays (DS and DR) have to be adjusted. Using a parameter set and the speed process, a simulated HR process can be simulated. By means of mathematical optimization, the optimal parameters can be determined in which simulated and original HR correspond best.

Simulations

After the individual calibration of the model to the athlete, it can be used for simulations. Putting in a speed progress, the user of the model gets out the corresponding HR process for the adapted athlete. In addition, the simulation provides the Reserve value, which provides an indication of nearness to the overloading mechanism:

$$\text{Reserve}[t] = 1 - SP[t] \tag{7}$$

If Reserve reaches zero or becomes negative, the overload mechanism is activated. This should be avoided during a competition. We utilize this value for the simulation of goal time. Therefore, simulations with different constant speed levels are made. Optimal running speed is set to the speed level of the simulation, where Reserve reaches zero right at the end of the marathon. Hence, the overload mechanism is avoided just in time, meaning the athlete has exhausted their reserves.

However, all these simulations are only based on a single graded incremental test and assume a flat track. If some conditions change from the day of the graded incremental test to the day of the competition or the running route has an elevation profile, simulations are no longer precise. Using elevation profile with PerPot is theoretically possible by performing a speed transformation before simulation (Endler & Friedrich, 2016). However, HR reaction on incline and decline is highly individual. Thus, it is not possible to transform the speed to a reference speed by a universal formula. The executed study with 19 participants only showed a general tendency to a third-order polynomial formula with incline and decline degrees as variable. The polynomial weights are highly individual and should be identified by a complex treatment. Furthermore, the investigation only considered incline and decline between -2% and +6%; inclines greater than +6% and declines less than -2% might have other characteristics.

Changing conditions can be addressed using an online optimization. Therefore, a feedback system was built using the Pegasos framework (Dobiasch & Baca, 2016). Feedback is generated based on online PerPot-Run simulations on the server with actual data collected by the mobile device. Actual speed and HR is sent via the mobile device to a server. Using this data, parameters of the model, e.g., delays and their progress, can be adapted based on the actual data. Simulations with the renewed parameter combinations result in an actual finish time. A built-in text-to-speech synthesizer of the mobile device can be used for converting feedback messages in text form into spoken messages. Athletes are informed about their estimated finish time and whether they should increase, decrease or maintain their current running speed. The actualization occurs frequently but, to avoid an overload of information, it should not be done too often, e.g., every 5 minutes. The usage of real-time data covers implicitly the form of the day and all general conditions such as temperature, including the cheering of spectators; beginners in particular tend to run faster when spectators are clapping or there is motivating music. This also leads to an increase in HR level, which might have serious consequences such as, e.g., an early breakdown as worst case. Both speed and HR are measured and, hence, included in the further determination and prescription.

Results

In this section, results of all introduced models are presented. The first model introduced in the section *Concepts and methods of modelling* was the Riegel formula. The formula was validated in several studies including, e.g., Vickers and Vertosick (2016). They investigated 2497 runners and identified that predicted marathon goal times from Riegel's formula are 10 minutes or more too fast for about half of the runners. Taking world records as basis of Riegel's formula into consideration, this is not surprising.

The advancement of Blythe and Király (2016) is based on a broader data set of British runners over several years. They state an average prediction error of 3.6 minutes on elite marathon runners (top 25% of the database). Their method was not evaluated in an own study, but rather on available data sets from the database itself.

The first individual models introduced in the section *Concepts and methods of modelling* were the lactate thresholds. Since these models use lactate values of athletes as individual parameters, big databases do not exist upon which determined and actual goal times in marathons can be compared. Therefore, costly investigations must be designed, which is not realistic for a broader group of participants. Furthermore, different models exist for the determination of the lactate threshold, but some of them differ significantly in their results. Thus, only few investigations were executed, e.g., for triathlon (Lorenzo, Minson, Babb & Halliwill, 2011). Only one investigation includes prediction of goal times for marathon (Roecker, Schotte, Niess, Horstmann & Dickhuth, 1998). Roecker et al. compared 166 prediction results of lactate diagnostics using the model of Dickhuth et al. (1999). They found a high Pearson correlation of 0.93 comparing predicted and real mean running speed during a marathon. Mclaughlin, Howley, Bassett, Thompson and Fitzhugh (2010) compared, among others, speed at the lactate threshold with goal time performed at a 16 km time trial. They found a high Pearson correlation for both values over the 17 participants of -0.907.

The own-developed system PerPot-Run, which was introduced in the section *Methodological experience*, was validated within the doctoral thesis of the author of this chapter (Endler, 2013). Eight participants competed in 33 races. Participants ran graded incremental tests within 1 week of the races. The investigation compared half-marathon and marathon goal times with the PerPot-Run predicted goal times. The mean difference of the depicted version was 3.62% ± 3.1%. Further investigations of Raschke (2015) and Schloss (2017) achieved similar results of 3.71% ± 1.37% and 4.34% ± 2.42%, respectively. As advancement, they included training data in different ways and improved these predictions in most cases. Other advancements, e.g., usage of incline and decline information, may improve the results (Endler & Friedrich, 2016). However, all advancements were only tested in a few self-tests and not within a broader investigation.

The live version of PerPot-Run was tested in 2015 in an unpublished investigation. Thirty-four athletes undertook a free time trial with individual distances as fast as possible. After a few days, they ran the same distance using the PerPot-Live system. Overall, 27 of 34 (79.4%) participants improved their performance by using the new feedback system. Two (5.9%) participants ran with identical average speed and five (14.7%) participants had a better performance without the system. The mean percentage speed improvement of the 27 participants who had an improvement was 7.59 ± 6.83%. Unfortunately, the split times were not analyzed.

This was done in the investigation of Pradier et al. (2016). After the first 5 km of the marathon, the model predicted the goal time with an average error of 3:47 minutes. This is a worse prediction than a similar investigation of Hammerling et al. (2013) who stated 2:49 minutes after running 5 km. However, after more than half the distance of the marathon, predictions of Pradier et al. became a little more precise (1:15 minutes) compared to Hammerling et al. (1:16 minutes).

Example

In the section *Methodological experience*, the three phases of HR reaction during a marathon were introduced. The first phase only concerns the temporary effect of a varying speed. If speed is increased, HR initially increases quickly. The increment, also referred to as increment rate, reduces until the HR does not increase anymore. The other way around, with decreasing speed, HR initially decreases quickly and the decrement rate reduces until the HR reaches its specific level for this speed level. This behavior can be described by an exponential delayed model. The discretized form is modeled as follows:

$$HR(t) = HR(t-1) + HR_{Rate} \tag{1}$$

with the (incremental/decremental) rate:

$$HR_{Rate} = (HR_{Goal} - HR(t-1)) * a * \Delta t \tag{2}$$

These easy formulas can be implemented using simple spreadsheet software such as Excel. At first, one must define some fixed cells for the model parameter:

- $HR(0)$: Specifies the starting heart rate.
- HR_{Goal}: Specifies the targeted end heart rate.
- a: Specifies the smoothing factor. This factor influences the delay until HR reaches the HR_{Goal}. $a \in \mathbb{R}$
- Δt: Specifies the time interval, i.e., how small the time steps of the simulated HR are, because in real life, HR cannot be measured more precisely than every second, "$t \in \mathbb{N}$.

In the next step, two formula columns must be implemented with the two formulas (1) and (2) above. Formula (1) first applies in the second row. The value in the first row is set by $HR(0)$.

Figure 9.4 shows the behavior for some typical smoothing factors between 0 and 1. The smaller a is, the longer it takes for HR to reach its final HR level.

The intended behavior of HR is only shown for $0 < a \leq 1$. For a = 0, nothing happens because the rate (2) is zero in every step. Figure 9.5 shows the behavior for $a < 0$ and $a > 1$.

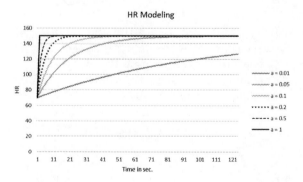

Figure 9.4 HR modelling with different smoothing factors between 0 and 1.

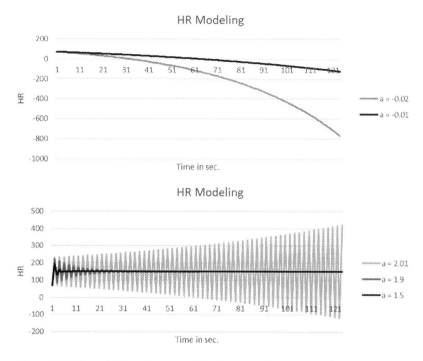

Figure 9.5 HR modelling using smoothing factors in the range a < 0 (top) to a >1 (bottom).

For $a < 0$, HR decreases in every step. This leads to a negative exponential decrease. The smaller a is, the faster HR decreases.

For $a > 1$, three similar but different behaviors can be seen. For $1 < a < 2$, HR oscillates around the final HR and stabilizes at that HR after a certain time. The smaller a, the faster that HR stabilizes. For $a = 2$, HR oscillates between the HR(0) and $2 * HR_{Goal} - HR(0)$. For $a > 2$, HR oscillates around the final HR but doesn't stabilize. HR recedes from the final HR in both directions, positively and negatively.

In summary, it can be stated that for HR modelling purposes, the smoothing factor must be constrained to $0 < a \leq 1$. Otherwise, our model shows no natural behavior of HR. Finding useful parameter values (or combinations) is one of the main challenges in modelling.

Exercise

The example above can be extended. Using an intermediate step, the HR progress is modeled a bit more realistically, thus:

$$HR(t) = HR(t-1) + HR_{Rate} \qquad (1)$$

$$x(t) = x(t-1) + x_{Rate} \qquad (2)$$

with the (incremental/decremental) rate:

$$HR_{Rate} = (x(t-1) - HR(t-1)) * a * \Delta t \qquad (3)$$

$$x_{Rate} = (HR_{Goal} - x(t-1)) * a * \Delta t \qquad (4)$$

Note: x(t) models the simple HR introduced in the previous paragraph.
Model parameters remain equal as described in the previous paragraph.
Exercise working steps:

1. Reserve four cells of a table spreadsheet (e.g., Excel) for the four model parameters HR(0), HR_{Goal}, a and Δt.
2. Implement the four formulas in four different columns. The formulas should start in the second row. The fixed value HR(0) of the first working step should be set as HR(0) value and x(0) value in the first row.
3. Plot HR and x values as a line chart.
4. Try out different parameter combinations, especially different smoothing factors. What kind of different behavior can you identify? Are parameter constraints evident again?

The example, as well as the exercise, were set as a task for 52 students during the lecture "Computer Science in Sports" during the 2017 summer term at Johannes Gutenberg University. The exercise was successfully solved by all of them within 1 week.

Note

1 www.thepowerof10.info/

References

Blythe, D. & Király, F. (2016). Prediction and quantification of individual athletic performance of runners. *PLoS One*, *11*(6), e0157257.

Brooke, J. & Hamley, E. (1972). The heart rate: Physical work curve analysis for the prediction of exhausting work ability. *Medicine and Science in Sports*, *4*(1), 23–26.

Bunc, V., Heller, J., Leso, J., Šprynová, Š. & Zdanovicz, R. (1987). Ventilatory threshold in various groups of highly trained athletes. *International Journal of Sports Medicine*, *8*(4), 275–280.

Dickhuth, H-H., Yin, L., Niess, A., Rocker, K., Mayer, F., Heitkamp, H. & Horstmann, T. (1999). Ventilatory, lactate-derived and catecholamine thresholds during incremental treadmill running: Relationship and reproducibility. *International Journal of Sports Medicine*, *20*(2), 122–127.

Dobiasch, M. & Baca, A. (2016). Pegasos: Ein Generator für Feedbacksysteme. In 11 Symposium der dvs Sportinformatik. Magdeburg: Otto von Guericke Universität.

El Helou, N., Tafflet, M., Berthelot, G., Tolaini, J., Marc, A., Guillaume, ... Toussaint, J-F. (2012). Impact of environmental parameters on marathon running performance. *PLoS One*, *7*(5), e37407.

Ely, M., Cheuvront, S., Roberts, W. & Montain, S. (2007). Impact of weather on marathon-running performance. *Medicine and Science in Sports and Exercise*, *39*(3), 487–493.

Endler S. (2013). *Adaptation of the meta model PerPot to endurance running for optimization of training and competition*, Doctoral dissertation. Johannes Gutenberg University, Mainz.

Endler, S. & Friedrich, O. (2016). Determining reference speeds and heart rates for endurance running prescription in altitude profile. In L. Lamas & T. Russomanno (Eds.), Proceedings of The 2016 International Association of Computer Science in Sport (IACSS) Conference, pp. 96–101. Brasilia. Campinas: FEF/UNICAMP.

Faude, O., Kindermann, W. & Meyer, T. (2009). Lactate threshold concepts: How valid are they? *Sports Medicine*, *39*(6), 469–490.

Ferley, D. & Vukovich, M. (2015). Time-to-fatigue during incline treadmill running: Implications for individualized training prescription. *Journal of Strength and Conditioning Research*, *29*(7), 1855–1862.

Hammerling D., Cefalu, M., Cisewski, J. Dominici, F., Parmigiani, G. & Paulson, C. (2014). Completing the results of the 2013 Boston Marathon. *PLoS One*, *9*(4), e93800.

Heaps, C., Gonzalez-Alonso, J. & Coyle, F. (1994). Hypohydration causes cardiovascular drift without reducing blood volume. *International Journal of Sports Medicine*, *15*(2), 74–79.

Kaikkonen, P., Hynynen, E., Mann, T., Rusko, H. & Nummela, A. (2010). Can HRV be used to evaluate training load in constant load exercises? *European Journal of Applied Physiology*, *108*(3), 435–442.

Keul, J., Simon, G., Berg, A., Dickhuth, H-H., Goerttler, I. & Kübel, R. (1979). Determination of the individual anaerobic threshold for performance assessment and training prescription. *Deutsche Zeitschrift für Sportmedizin*, *30*, 212–218.

Laborde, S., Dosseville, F. & Allen, M. (2016). Emotional intelligence in sport and exercise: A systematic review. *Scandinavian Journal of Medicine and Science in Sports*, *26*, 862–874.

Lorenzo, S., Minson, C., Babb, T. & Halliwill, J. (2011). Lactate threshold predicting time-trial performance: Impact of heat and acclimation. *Journal of Applied Physiology*, *111*(1), 221–227.

Mclaughlin, J., Howley, E., Bassett, D., Thompson, D. & Fitzhugh, E. (2010). Test of the classic model for predicting endurance running performance. *Medicine & Science in Sports & Exercise*, *42*(5), 991–997.

Parak, J. & Korhonen, I. (2014). Evaluation of wearable consumer heart rate monitors based on photopletysmography. *Engineering in Medicine and Biology Society (EMBC): 36th Annual International Conference of the IEEE*, 3670–3673.

Perl, J. (2004). PerPot: A meta-model and software tool for analysis and optimisation of load-performance-interaction. *International Journal of Performance Analysis of Sport*, *4*(2), 61–73.

Péronnet, F., Thibault, G., Rhodes, E. & McKenzie, D. (1987). Correlation between ventilatory threshold and endurance capability in marathon runners. *Medicine and Science in Sports and Exercise*, *19*(6), 610–615.

Pradier, M., Ruiz J. & Perez-Cruz, F. (2016). Prior design for dependent dirichlet processes: An application to marathon modeling. *PLoS One*, *11*(1), e0147402.

Raschke, P. (2015). Evaluation der Zielzeitoptimierung im ausdauerorientierten Laufsport mittels des sportinformatischen Modells PerPot im Halbmarathon/

Marathon. Diploma thesis, Johannes Gutenberg University Mainz, Institute of Sport Science.

Reybrouck, T., Ghesquiere, J., Weymans, M. & Amery, A. (1986). Ventilatory threshold measurement to evaluate maximal endurance performance. *International Journal of Sports Medicine*, 7(1), 26–29.

Riegel, P. (1981). Athletic records and human endurance: A time-vs.-distance equation describing world-record performances may be used to compare the relative endurance capabilities of various groups of people. *American Scientist*, 69(3), 285–290.

Roecker, K., Schotte, O., Niess, A., Horstmann, T. & Dickhuth, H-H. (1998). Predicting competition performance in long-distance running by means of a treadmill test. *Medicine and Science in Sports and Exercise*, 30(10), 1552–1557.

Schloss, A. (2017). Evaluation of optimization of target time in endurance oriented running using the sport informatic model PerPot. Master's thesis, Johannes Gutenberg University Mainz, Institute of Sport Science.

Stevinson, C. & Biddle, S. (1998). Cognitive orientations in marathon running and hitting the wall. *British Journal of Sports Medicine*, 32, 229–235.

Svendahl, K. & MacIntosh, B. (2003). Anaerobic threshold: The concept and methods of measurement. *Canadian Journal of Applied Physiology*, 28(2), 299–323.

Varley, M., Fairweather, I. & Aughey, R. (2012). Validity and reliability of GPS for measuring instantaneous velocity during acceleration, deceleration, and constant motion. *Journal of Sports Sciences*, 30(2), 121–127.

Vergès, S., Flore, P. & Favre-Juvin, A. (2003). Blood lactate concentration/heart rate relationship: Laboratory running test vs field roller skiing test. *International Journal of Sport Medicine*, 24(6), 446–451.

Vickers A. & Vertosick, E. (2016). An empirical study of race times in recreational endurance runners. *BMC Sports Science, Medicine and Rehabilitation*, 8(1), 26.

10 Training

Christian Rasche and Mark Pfeiffer

Introduction

Training is mostly targeted on achieving training goals that, in turn, result in physical adaptation and ultimately comply with superior intents, such as increasing performance capability, gaining fitness, losing weight and many more. Thus, the relationship between training and performance is of particular significance for many applications of sport, such as rehabilitation sports, recreational sports and especially competitive sports (McArdle, Katch & Katch, 2010; Soligard et al., 2016). Since improving or stabilizing athletic performance is a particular goal in competitive sports, the effect of training on performance is of vital interest to coaches organizing and scheduling their athletes' training programs. Therefore, it is important to learn the mechanics and basics of how training load transfers to adaptation processes and ultimately affects performance.

Performance in sports is closely connected to physiological adaptations that are induced by the athlete's training program. However, physiological adaptation is a complex, nonlinear problem and the inter- and intraindividual variability in both, the physiological response to training as well as the relationship between training adaptations and performance, is huge (Borresen & Lambert, 2009). A repeated training stimulus may provoke varying effects on an athlete because the adaptation of a biological system itself entails modified subsequent adaptations. Furthermore, training-induced adaptations are vastly dependent on a framework of factors related to human genetics, training status and history, psychological factors and many more, in which each one has an impact on the capacity to respond to training. In addition to the extensive body of knowledge on the basic physiological mechanisms of adaptation, there are different approaches to analyzing long-term adaptation caused by training at regular intervals, especially with regard to effects on performance (Clarke & Skiba, 2013; Taha & Thomas, 2003).

Project/problem

Recently, a number of attempts have been made to model training effects on performance, which may be labeled as *performance modelling* or *training effect analysis*, depending on the point of view. The underlying methodical approach is the simulation of athletic performance development triggered by training using concepts and modelling assumptions based on the underlying physiology and mechanisms. In general, similar training stimuli do not lead to similar adaptive reactions, neither in one person nor among athletes. Hence, adaptation processes are characterized by a highly inter- and intraindividual variability that has to be reflected in the models' structures. Consequently, the approaches presented in the following make use of single-case time series and the athlete represents a dynamic system with training load as the input and performance as the output to model the nonlinear training-performance relationship.

Concepts and methods of modelling

To deal with the above-mentioned problems and challenges, scientists have employed different mathematical modelling techniques and concepts. This subsection gives an overview of the existing methodical approaches by summarizing their similarities on the one hand and contrasting them based on underlying physiological concepts on the other hand. The specific models are termed *performance models*.[1]

In general, all performance models employ intraindividual (single-case) time series with the quantification of training loads being considered the independent input variables and performance the single dependent output variable. Regarding the training-performance relationship, there is a huge amount of research and several findings, which may be incorporated into a modelling approach. Consequently, performance models are characterized by incorporated *physiological assumptions* that determine the model's technical and mathematical structures. Each model emphasizes particular qualities to handle the specificity of training and adaptation, such as causality and predictability, interdependencies and nonlinearity, number of training variables and physiological phenomena. Each one has its own strength and weaknesses, while the appropriate model depends on the context, research aim and available resources. On the basis of a reasonable framework, sport scientists may learn about individual adaptation processes and transfer results easily into sport practice.

This subsection focuses on five specific physiological assumptions that are derived from the literature and reflected by performance models (see Figure 10.1). The assumptions overlap in such a way that the first one is applicable for all performance models, while each consecutive assumption decreases the number of includable performance models.

Two physiological assumptions concern *time dependencies* of the comprised variables and mainly characterize performance models based on statistical analysis and machine-learning algorithms. Three physiological assumptions implement basic knowledge of exercise physiology, merged as *training effects*, and embrace a number of performance models that may be divided into *type A* and *type B* performance models.

All physiological assumptions and their motivation are listed and explained below. Additionally, a comprehensive overview of physiological assumptions, methods and performance models is given in Figure 10.1.

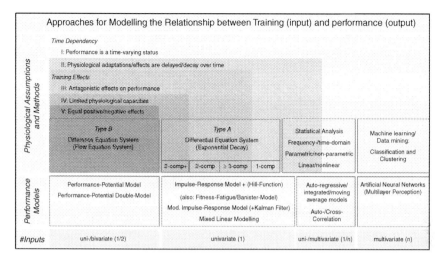

Figure 10.1 Overview of existing performance models and their physiological assumptions, methods and number of inputs.

Time dependency

Performance is a time-varying status

The most general physiological assumption is a time-varying status of performance. Without further restricting the underlying physiological relationships of training and performance, there are some performance modelling approaches located in machine-learning research fields using artificial neural networks (Edelmann-Nusser, Hohmann & Henneberg, 2002; Haar, 2011; Hohmann, Edelmann-Nusser & Henneberg, 2000). Even though these nonlinear, supervised learning approaches are becoming more popular (Pfeiffer & Hohmann, 2012), one may criticize them for not containing enough assumptions about the relationship between training and performance to make good use of resulting simulations and predictions. Their practical use is limited due to the common black-box character of machine-learning algorithms and their inability to retransfer knowledge about training arrangements.

Physiological adaptations/effects are delayed/decay over time

The second assumption characterizing a number of approaches presumes that physiological adaptation processes triggered by training vary over time regarding their impact on performance. Hence, the athletes' performance is influenced by training with changing magnitudes at different points of time, causing alterations on the initial performance.

There are some classical statistical regression approaches (e.g., cross-correlation) that are based on delayed dose-response characteristics (Collette, Pfeiffer & Kellmann, 2016; Ferger, 2010; Osterburg, Rojas, Strüder & Mester, 2002, 2003), but performance models located in the research fields of statistical analysis and machine-learning algorithms do not incorporate further assumptions about specific physiological effects of training on performance. To stay inside a reasonable framework based on research findings, this subsection focuses on performance models incorporating physiological assumptions regarding training effects (type A and type B performance models; see Figure 10.1).

Training effects

Antagonistic structure

Regarding adaptations, human physiology may be characterized by opposing effects of training on body structures (Hausswirth & Mujika, 2013; McArdle et al., 2010). Consequently, *antagonistic* performance models assume that athletic training has concurrent effects. On the one hand, physical activity strains the human body, resulting in fatigue (the negative effect of training) and, on the other hand, it aims to improve performance by increasing fitness (the positive effect of training).

Banister, Calvert, Savage and Bach (1975) and Calvert, Banister, Savage and Bach (1976) proposed the first performance modelling idea, which features physiologically motivated antagonistic training effects. Calvert et al. (1976) illustrate a multicomponent model that includes the performance determinants "endurance" (cardiovascular), "strength," "skill" and "psychological factors" and various cross-effects among them. By applying it to a swimmer, they end up with a time function of fitness and fatigue. In the research, the concept of different factors affecting performance positively and negatively led to the general formulation of so-called *components* (Busso, Carasso & Lacour, 1991; Morton, Fitz-Clarke & Banister, 1990). A component stands for a single effect—either positive or negative—that connects training and performance. Following these denominations, type A performance models can be differentiated by the number of components included, ranging from one component (1-comp) to three components or more (≥3-comp).

Initially, Calvert, et al. (1976) based their number of three components on the model's interpretation and imitation of actual physiological processes based on the literature, and Morton et al. (1990) reformulated the model using two components. Anyway, Busso et al. (1991) modeled data sets using a variable number of components and compared the resulting model-fit to analyze how many components provide the best statistical adequacy. Based on their findings, they reasoned that a 2-comp model provides a proper representation of the training process, and most researchers adhere with the recommendation of a 2-comp type A performance model (Banister, Carter & Zarkadas, 1999; Banister, Morton & Fitz-Clarke, 1992; Chalencon et al., 2012). Other researches followed this example and conducted studies optimizing the model-fit of given data by varying the included components besides optimizing the model's parameters (Busso et al., 1992; Millet et al., 2002; Millet, Groslambert, Barbier, Rouillon & Candau, 2005; Morton, 1991; Philippe et al., 2015). Only a few studies employ type A performance models with a single positive decaying effect of training (Agostinho et al., 2015; Candau, Busso & Lacour, 1992). Type B performance models and most of the type A performance models are based on antagonistic effects of training on performance.

Limited physiological capacities

Recent research about functional and non-functional overreaching and overtraining (Halson, 2014; Schwellnus et al., 2016; Soligard et al., 2016) as well as basic exercise physiology justify the assumption that physiological capacities are limited. The finiteness of natural capabilities as well as the inability of human physiology to improve performance endlessly is taken into account by either defining individual, maximal values of fitness and fatigue or incorporating feedback mechanisms to take care of accumulated fatigue that results in performance decrements. Type B performance models and some type A performance models incorporate limited physiological capacities (see *Examples of type A performance models*).

Equal positive/negative effects

Human physiology is structurally characterized by a homeostatic behavior concerning the training-performance relationship (Hausswirth & Mujika, 2013; McArdle et al., 2010). Classically, performance is increased by training and regresses to the former base level absent of training loads. Accordingly, the last physiological assumption narrowing the model's structures states that positive and negative training effects are commensurate with each other, i.e., training affects fitness and fatigue to the same extent. To date, this applies only for type B performance models.

Methodological experience

The first part of this subsection will deal with type A and type B *antagonistic* performance models in more detail. Both types and their characteristics will be described separately to illustrate their key mechanics and technical implementations. A comprehensive comparison is given in the second part to point out the most important differences for further research and practical implementation.

The descriptions of type A and type B performance models are structured threefold. The first part, (i) performance synthesis and alteration, treats the computation of modeled performance based on diverging principles as well as different ways of performance alterations by delaying and decaying mechanisms.

Consequently, type A and type B performance models include specific parameters[2] referring to fitness and fatigue respectively to take account of individual physiological dynamics. Second, some performance models restrict the possible effects of the processed training data on fitness, fatigue and, finally, performance, resulting in (ii) presumptions about fitness, fatigue and performance constraints. Third, since training and performance usually originate from different scales, a (iii) compensation of diverging measurement units of input and output is needed by means of scaling factors.[3] In general, scaling factors are arbitrary units and do not have a physiological equivalent, but there are important differences regarding their interpretation in the models' contexts.

Type A performance models

(i) Performance synthesis and alteration

Type A performance models are predicated on first-order differential equations, which are solved by exponential decay functions (see *Examples of type A performance models* for technical details). The consolidation of positive and negative effects towards performance $p(t)$ is realized as a sum of the base level of performance p_0 with fitness and fatigue components and is called *transfer function* (Equation 1).

$$p(t) = p_0 + fitness(t) - fatigue(t) \qquad (1)$$

Both training effects—positive and negative—are initially fed equally by training loads and decay depending on their specific time constants. Notably, while a decay of the accumulated fitness decreases performance, decay of fatigue increases performance.

Furthermore, without any additional model implementations, constant training loads lead to a transient oscillation towards a lower performance level if the positive time constant is smaller than the negative one, and towards a higher performance level if the time constant ratio is the other way around. The total variation is reliant on the training loads' magnitude. Performance returns to the starting level (p_0) absent of training loads, because fitness and fatigue decay completely in a period determined by the respective time constants (see *Comparison of type A and type B performance models*).

(ii) Presumptions about fitness, fatigue and performance constraints

Most of the type A performance models do not limit the accumulation of fitness and fatigue. Therefore, in theory, huge amounts of training load are modeled without considering physiological limits resulting in performance increments above reasonable boundaries. However, there are modifications presented in the literature to consider cumulated fatigue or restrict the effective load (Busso, 2003; Hellard et al., 2005), termed *2-comp+* in Figure 10.1. By amplifying the negative effect of training or limiting training loads mathematically, unrealistic performance increases emanating from excessive training loads are prevented, which is shown in *Examples of type A performance models*.

(iii) Compensation of diverging measurement units

The implementation of scaling factors to fit training and performance measurements is realized for accumulated fitness and fatigue levels separately, while performance measurements remain at their original level. Notably, different scaling factors imply unequal emphasis of fitness and fatigue effects towards performance.

Examples of type A performance models

The most popular example of a type A model is the Impulse-Response Model (IR Model)[4] first published in 1975 by Banister and colleagues (Banister et al., 1975; Calvert et al., 1976) and enhanced to its most used form by Morton et al. (1990) and Busso et al. (1991). Since then, it has been applied numerous times using different notations and extensions and will be presented in the following.

Based on the idea that training responses follow first-order kinetics, the basic framework comprises a set of two solutions of first-order differential equations. The resulting exponential functions for fitness and fatigue implement the decaying effects of training and are integrated using the aforementioned transfer function (Equation 1) towards a discretized sum formula for daily performance computation (Equation 2).

$$p(t) = p_0 + k_1 \cdot \underbrace{\sum_{i=1}^{t-1} w(i) \cdot e^{\frac{-(t-i)}{\tau_1}}}_{Fitness} - k_2 \cdot \underbrace{\sum_{i=1}^{t-1} w(i) \cdot e^{\frac{-(t-i)}{\tau_2}}}_{Fatigue} \qquad (2)$$

A comprehensive derivation including mathematical details can be found in Clarke and Skiba (2013). Modeled performance on a given day is computed either before (preload) or after (postload) daily training. Since performance in the morning is of interest concerning competitions or scheduling issues, the previously introduced Equation (2) results in a preload performance. Performance $p(t)$ is computed given past training loads $w(i)$, which are decayed daily by factors between 0 and 1 ($0 < e^{\frac{-(t-i)}{\tau_1}} \leq 1$) depending on individual time constants τ_1 and τ_2 and scaled by k_1 and k_2 regarding fitness and fatigue respectively. Equations (3) to (7) represent the same model using recursive equations and their aggregation towards a preload performance.

$$g(t) = g(t-1) \cdot e^{\frac{-1}{\tau_1}} + w(t) \qquad (3)$$

$$h(t) = h(t-1) \cdot e^{\frac{-1}{\tau_2}} + w(t) \qquad (4)$$

$$fitness(t) = k_1 \cdot (g(t) - w(t)) \tag{5}$$

$$fatigue(t) = k_2 \cdot (h(t) - w(t)) \tag{6}$$

$$p(t) = p_0 + fitness(t) - fatigue(t) \tag{7}$$

Busso (2003) published a modification for the IR Model, termed the modified IR Model[5] (IRmod Model), to cover changes of magnitude and duration of fatigue induced by training, by adding a gain term $k_3(i)$ resulting in an additional time constant variable τ_3 (equations 8 and 9).

$$p(t) = p_0 + k_1 \cdot \underbrace{\sum_{i=1}^{t-1} w(i) \cdot e^{\frac{-(t-i)}{\tau_1}}}_{Fitness} - k_2 \cdot \underbrace{\sum_{i=1}^{t-1} k_3(i) \cdot w(i) \cdot e^{\frac{-(t-i)}{\tau_2}}}_{Fatigue} \tag{8}$$

$$\text{and } k_3(i) = \underbrace{\sum_{j=1}^{i} w(i) \cdot e^{\frac{-(i-j)}{\tau_3}}}_{Gain\ Term} \tag{9}$$

The gain term increases fatigue in response to excessive loads and, therefore, can be interpreted as a limitation of physiological capacities. The recursive equations for the IRmod Model to compute the preload performance can be found below (equations 10–15).

$$g(t) = g(t-1) \cdot e^{\frac{-1}{\tau_1}} + w(t) \tag{10}$$

$$h(t) = (h(t-1) + w(t-1) \cdot \dot{h}(t-1)) \cdot e^{\frac{-1}{\tau_2}} \tag{11}$$

$$\dot{h}(t) = \dot{h}(t-1) \cdot e^{\frac{-1}{\tau_3}} + w(t) \tag{12}$$

$$fitness(t) = k_1 \cdot (g(t) - w(t)) \tag{13}$$

$$fatigue(t) = k_2 \cdot h(t) \tag{14}$$

$$p(t) = p_0 + fitness(t) - fatigue(t) \tag{15}$$

Furthermore, Hellard et al. (2005) adapted the implementation of training loads $w(i)$ by using a Hill function to limit their effects on performance. Avalos, Hellard and Chatard (2003) complemented the 2-comp IR Model by a mixed linear modelling approach, while Matabuena and Rodriguez (2016) extended it towards a delay differential model in which previous training sessions are accounted for inside the recursion equations. Kolossa et al. (2017) extended the IRmod Model (Busso, 2003) by using a Kalman Filter to consider measurements of performance to adapt the fitness and fatigue estimates.

Type B performance models

(i) Performance synthesis and alteration

Type B performance models access difference equations (recurrence relations) to realize flows between three so-called potentials: fitness, fatigue and performance. Fitness and fatigue potential levels discharge into the performance potential by means of stepwise flows determined by specific delays (Figure 10.2).

A flow of the fitness or fatigue potential towards the performance potential results in a performance increase or decrease respectively. Furthermore, without any additional model implementations, constant training loads lead to a transient oscillation towards a higher performance level if the positive delay is smaller than the negative delay, and towards a lower performance level if the delay ratio is the other way around. The total variation is reliant on the training loads' magnitude as well as scaling factors and min/max operators (see below). Due to the equal feed of the fitness and fatigue potential by means of training load, performance returns to the starting level absent of training loads because potentials deplete in a space of time determined by the delays.

(ii) Presumptions about fitness, fatigue and performance constraints

Type B performance models feature fixed constraints for the fitness, fatigue and performance potential to account for the finiteness of natural capabilities as well as the inability of human physiology to improve performance endlessly. Accordingly, training overload is specified by an overfilling of the fatigue potential leading to additional negative flow towards the performance potential. Hence, flows between limited potentials have to be limited themselves by considering

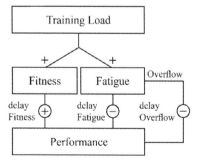

Figure 10.2 Type B performance model.
Source: Conceptually modified version of Perl's (2001) "Performance-Potential Model"

current potential levels and the available space to avoid a violation of determined model structures. By means of minimum and maximum operators, flows between potentials are forced to comply with potential boundaries and inherit nonlinear model behavior.

(iii) Compensation of diverging measurement units

The compensation of measurement units for type B performance models is carried out before any further procedure. Training load (input) as well as performance measurements (output) are scaled linearly towards a standardized range of 0 to 1. Separate factors for input and output are used, termed scaling factor training and performance (SFT/SFP), while the scaled training load is fed equally into the fitness and fatigue potentials. The premature determination of training and performance scaling factors defines quantified relationships between training and performance because it enables estimation of the magnitude a training session has on the athletes' performance as well as the individually possible performance increase. In other words, limited fitness, fatigue and performance-potential capacities from 0 to 1, in combination with a given scaling factor, reveal a presumption about individual physiological capacities of load and adaptation that is the exact magnitude of a given training load and the maximum impact training potentially has on performance.

Examples of type B performance models

The first type B performance model, the Performance-Potential Model (PerPot Model), was published by Mester and Perl (2000) and Perl (2001).

At first, all training and performance measurements are scaled towards a fixed range of 0 to 1 by using scaling factors SFT and SFP (equations 16 and 17).

$$training\ load_{scaled} = training\ load_{original} \cdot \frac{SFT}{\max(training\ load)} \quad (16)$$

$$performance_{scaled} = performance_{original} \cdot \frac{SFP}{\max(performance)} \quad (17)$$

Then, throughout the modelling process, daily training load is added equally to fitness and fatigue potentials every discrete time step (equations 18 and 19).

$$Fitness_{new} := Fitness_{old} + training\ load_{scaled} \quad (18)$$

$$Fatigue_{new} := Fatigue_{old} + training\ load_{scaled} \quad (19)$$

The flows towards the performance potential, termed *FitnessFlow*, *FatigueFlow* and *Overflow*, are determined by the delaying parameters *delayFitness*, *delayFatigue* and *delayOverflow* respectively as well as minimum and maximum operators (equations 20–22).

$$FitnessFlow := \frac{\min(\min(1, Fitness), \min(1, 1 - Performance))}{delayFitness} \qquad (20)$$

$$FatigueFlow := \frac{\min(\min(1, Fatigue), \max(0, Performance))}{delayFatigue} \qquad (21)$$

$$Overflow := \frac{\max(0, Fatigue - 1)}{delayOverflow} \qquad (22)$$

Consequently, fitness and fatigue potentials are upgraded accordingly (equations 23 and 24).

$$Fitness_{new} := Fitness_{old} - FitnessFlow \qquad (23)$$

$$Fatigue_{new} := Fatigue_{old} - FatigueFlow - Overflow \qquad (24)$$

Finally, the performance potential is changed referring to the physiological mechanisms (Equation 25).

$$\begin{aligned} Performance_{new} := {} & Performance_{old} + FitnessFlow \\ & - FatigueFlow - Overflow \end{aligned} \qquad (25)$$

Referring to Taha and Thomas (2003) that vital information of the training process is lost by reducing the training input to one number, Perl and Pfeiffer (2011) developed a model with two inputs based on the same principles as the PerPot Model. This model, the Performance-Potential Double Model (PerPot DoMo), allows to examine training processes by separating training volume and training intensity, which is crucial in endurance sports and especially in strength training (Kraemer & Fleck, 2007).

Comparison of type A and type B performance models

In the following, the vocabulary of type A and B performance models is adjusted slightly to simplify the comparison. Regardless of the specific model type, fitness and fatigue can be understood as collective states of all types of positive and negative effects. Therefore, both may be termed as *potentials* that contain a quantified amount of contents varying as a function of time. Furthermore, since time constants (type A) and delays (type B) are compatible, both are referred to as *delays* to omit unnecessary doublings.

Regarding performance synthesis and alteration, a number of individual characteristics differ as follows. In the case of type A performance models, modeled performance is the difference of current states of fitness and fatigue potentials itself determined by the transfer function (Equation 1), whereas modeled performance is altered stepwise by fitness and fatigue flows in type B performance models, with performance representing a third, dependent potential. Accordingly, a decaying fitness potential in type A models decreases performance, while a flow from the fitness towards the performance potential increases performance and vice versa in type B ones. The same applies to decaying and delayed fatigue. Therefore, decaying potentials in type A models affect performance *inversely* compared to delayed flows between potentials in type B models.

A closer look at the computation of decaying and delayed rates reveals that the *inverse performance development* deviates only by a small margin and is almost equivalent for delays bigger than six, which is illustrated in the following. In type A and B performance models, fitness and fatigue potentials are decayed and delayed respectively based on equations 3 and 4 and 20 to 24 as follows.

$$\text{Type A} \rightarrow potential_{new} = potential_{old} \times \underbrace{e^{\frac{-1}{delay}}}_{0 < decay\ factor < 1} \qquad (26)$$

$$\text{Type B} \rightarrow potential_{new} = potential_{old} - (potential_{old} \div delay)$$
$$= potential_{old} \times \underbrace{\frac{delay - 1}{delay}}_{0 \le delay\ factor < 1} \qquad (27)$$

Exemplary results of decay/delay factors and their deviation depending on delay values are shown in Table 10.1.

The deviation of the decay/delay factors decreases with rising delays and converges towards 0. Additionally, repeated application of decay/delay factors on a single training load leads to a general behavior as follows. After a period equal to the respective delay, the training load is decayed inside the potentials towards $e^{-1} \triangleq 0.37$ for type A performance models and towards $\left(\frac{n-1}{n}\right)^{n-1}$ of its initial value for type B performance models, whereas the latter converges towards e^{-1}.

Table 10.1 Exemplary delay values, resulting decay/delay factors for type A and B performance models and decay vs. delay factor % deviations.

	*delay value**						
	1	2	3	4	5	6	>6
Decay factor (type A): $\frac{-1}{e^{delay}}$	0.37	0.61	0.72	0.78	0.82	0.85	≥0.87
Delay factor (type B): $\frac{delay-1}{delay}$	0	0.50	0.66	0.75	0.80	0.83	≥0.86
Decay vs. delay factor: % deviation	–	21.3%	7.48%	3.84%	2.34%	1.58%	≤1.14%

*type A performance models: τ_1, τ_2, τ_3; type B performance models: *delayFitness, delayFatigue*

The next paragraph combines information on *presumptions about fitness, fatigue and performance constraints* with the *compensation of diverging measurement units* to highlight further differences and similarities between type A and B performance models.

From a physiological and practical point of view, it is import to estimate *how impactful* a training session is and *how long* it affects the athlete. Type A and B performance models address this challenge differently and, until now, have not found a comprehensive solution, although there are some indications for purposeful interpretation of the type B performance models' parameters. Physiological constraints of fitness and fatigue in combination with scaling factors may reveal the impact of a training session. Furthermore, a single training load is effective in fitness and fatigue potentials to a certain extent for a fixed time frame, based on the specific delays in type A and type B performance models. Dependent on an assumption—how much "training effect" of a training load remains in a potential, e.g., 1, 2, 5 or 10%—the time frame of depletion is computable. Finally, by simulating performance development by means of training load, inferences about the physiological assumptions and their validity as well as the verification of theories and findings in complex, long-term training processes may be made.

Optimization of model parameters

In practice, model parameters are of paramount importance because they determine the model's behavior, strengths and weaknesses. Anyway, there are overfitting and interdependency issues (Hellard et al., 2006) as well as unclear parameter constraints to address.

To compute the individual-specific parameters for type A and B performance models, the model-fit[6] between empirical and simulated performances is optimized by searching for appropriate parameter combinations. Table 10.2 presents the parameters being optimized for specific examples of type A and B performance models.

There are several approaches to assess the model-fit based on different concepts like total or relative deviation (e.g., residual sum of squares (RSS)); root-mean-square error (RMSE)); percentage errors (e.g., mean average percentage error (MAPE)); correlation measures (e.g., Pearson's r, coefficient of determination (r^2)); intraclass correlation coefficient (ICC) and other specific indices, such as

Table 10.2 Included parameters of selected type A and B performance models in the optimization process (τ_1, τ_2, τ_3, delayFitness, delayFatigue = delays and k_1, k_2, SFT, SFP = scaling factors).

Performance Model	Parameters
IR Model	τ_1, τ_2, k_1, k_2
IRmod Model	$\tau_1, \tau_2, \tau_3, k_1, k_2$
PerPot Model	$delayFitness, delayFatigue, SFT, SFP$

the Akaike Information Criterion (AIC). Based on optimization algorithms and methods (e.g., combinatorial optimization, evolutionary algorithms or gradient methods) the minimum/maximum of a given model-fit criteria, including the corresponding parameters, is determined. Commonly, performance models' optimization algorithms minimize the RSS, while the first empirical performance measurement is mostly used as the basic level of performance and represents the starting point of any performance alteration. Sometimes, the starting performance is optimized besides the other parameters, especially if there are few empirical performance measurements in the beginning of a data set. To emphasize recent performance residuals more than older ones, recursive least squares algorithms may be employed. Busso, Denis, Bonnefoy, Geyssant and Lacour (1997) refined their optimization in a time-variant approach of the IR Model further by estimating the optimal parameter set each time data was collected.

Data requirements and treatment

To maximize the efficiency of training effect analysis, training (input) and performance (output) data should comply with basic requirements. First, the performance criteria of data collection are of major importance, namely, objectivity, reliability and validity. If the aim of performance modelling is to learn about individual adaptation processes, the processed data must yield quality standards and contain relevant information with little measurement error. Furthermore, closely meshed measurements are required since performance models are based on daily computations. Missing data denies important information that is needed for continuous modelling, and may lead to non-interpretable model behavior.

Even though a continuous computation of performance levels is possible, performance models are commonly used with discrete time steps using natural numbers, mostly representing days or weeks, because these are the most important and common periods for training scheduling and regeneration. Furthermore, the practical application of a specific performance model in sports science mostly consists of two steps: First, the models' parameters are optimized using a certain time frame of the existing data. Second, a performance prediction is conducted for a subsequent time frame to assess the predictive accuracy.

Results

This subsection features selected international research studies to illustrate the principles and areas of application of performance modelling as well as give an overview regarding research emphasis and review articles. The subordinate research question is about the relationship of training and performance, while specific aims include the application of taper phases, choice of input variables, model-derived information such as physiological correlates for fitness and fatigue as well as sport-specific applications. The presented research comprises mainly type A and B performance models, given that most research has been conducted for type A performance models since they were published 25 years earlier than type B performance models and other approaches are less popular.

The existing reviews comprise structural and mathematical aspects (Hellard et al., 2006) as well as summaries and comparisons of research results (Pfeiffer, 2008; Taha & Thomas, 2003). Additionally, information about the quantification of training load and performance (Jobson, Passfield, Atkinson, Barton & Scarf, 2009) and resources for the application of performance models in practice (Clarke & Skiba, 2013) is embedded.

Performance model-specific research has mainly been conducted in the 1990s and early 2000s. In 1991, Fitz-Clarke, Morton and Banister published the concept of influence curves, derived from the IR Model. Influence curves provide information about the time-dependent effects of single training loads on performance and generate model-derived information on the time to recover performance t_n (influence curve turns from negative to positive values) and the time to peak performance t_g (maximum of the influence curve) using the model parameters τ_1, τ_2, k_1 and k_2. These computations are restricted to 2-comp IR Models, since other types of type A performance models (1-comp, 2-comp+ and ≥3-comp) contain different model parameters. In practice, the information is used to assess ideal taper phases and prepare optimally for upcoming tournaments and competitions (Mujika et al., 1996).

Several research groups have investigated physiological correlates for fitness and fatigue to validate the type A performance modelling concept. Morton (1997) summarized the results of studies concerning iron status (Banister & Hamilton, 1985), hormonal responses (Busso et al., 1990; Busso et al., 1992), enzyme activities (Banister et al., 1992) and urea, serum protein and others (Banister & Fitz-Clarke, 1993) stating that there are basic similarities, but no perfect phase. However, the existence of positive correlations between physiological correlates and fitness as well as fatigue impedes the model's interpretation (Taha & Thomas, 2003) and suggests that there are no perfectly fitting physiological equivalents for the 2-comp IR Model.

Millet et al. (2002) used type A performance models to examine relationships and interdependencies between load and performance in triathlon. They separated swimming, running and cycling loads and modeled them independently towards performance respectively, using one to three components. The relationships between load and performance were significant for running and swimming, and cross-training effects were shown for cycling and running, but not for swimming.

To compare the effect of different training loads (input variables), Wallace, Slattery, and Coutts (2014) employed session-RPE, TRIMP and rTSS (Halson, 2014) as training quantification in endurance athletes and modeled them on performance quantified by submaximal heart rate and heart-rate variability (HRV) indices individually. By using the IRmod Model (2-comp+), they assessed the correlations of actual and modeled performance for each combination of training and performance (input and output) variables and discovered that each training quantification is equally appropriate for modelling actual training responses.

Since the publication of type B performance models in the early 2000s, some research groups examined and compared type A and B performance models. Ganter, Witte and Edelmann-Nusser (2006) compared the IR- and the PerPot Model regarding their performance prediction following an 8-week cycling training of 10 sport students. Model-fits were assessed by the coefficient of determination and the MAPE were superior for the IR Model. Anyway, the authors prefer the PerPot Model, arguing that physiological interpretation of model parameter remains questionable for the IR Model. Pfeiffer (2008) compared the IR- and PerPot Model regarding their model-fit and predictive accuracy. Six untrained cyclists performed two studies consisting of 7 and 10 weeks of training on bicycle ergometers. The resulting training load was modeled to an all-out test, which was executed on a bicycle ergometer too. Results showed that the PerPot Model achieved better model-fits for 9 out of 15 cases regarding modelling (calibration phase) and on average for predictions, while individual prediction was superior in 8 out of 15 subjects for the IR Model.

Performance modelling remains an important task to validate actual physiological processes and gain knowledge about the relationship of training and performance. However, a reasonable choice of training and performance quantification based on research findings is of huge importance to justify the assignment of performance models and to assume a valid input-output relationship.

Example

This subsection will give an example of performance modelling that demonstrates the modelling idea and how it works for the IRmod Model (antagonistic type A performance model; 2-comp+).

First, available training and performance data has to be preprocessed by checking for artifacts and creating equal time series for each, resulting in a $(n \times 2)$-vector,

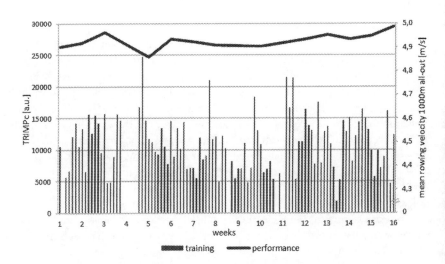

Figure 10.3 15 weeks of training (gray bars, TRIMPc [a.u]) and performance (gray line, mean rowing velocity 1000m all-out [m/s]) measurements for an elite rower.

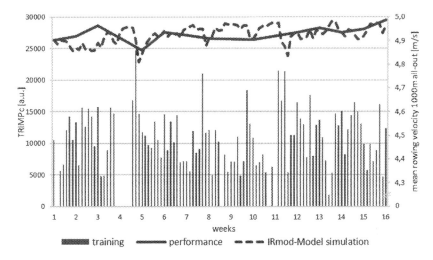

Figure 10.4 Performance simulation (dashed line) with the IRmod Model.

where n are the days. Multiple training sessions are accumulated and treated as one daily value. If there was no training conducted or performance measurement done, the daily value is 0 in both cases. Then, either all data or coherent parts of the vector may be used for the optimization process depending on the aim of the analysis.

In Figure 10.3, real data for an elite rower over a 3-month period is shown, whereas the complete data set is used for the modelling process.

The first empirical performance measurement is set as starting performance p_0 and equations 10–15 are used recursively to compute the simulated performance by processing the daily training load $w(t)$. This process is repeated with specific parameter sets ($\tau_1, \tau_2, \tau_3, k_1, k_2$) based on the optimization process, until the RSS of simulated and empirical data is minimized. The corresponding parameter set for the data above is $(20, 1, 1, 4.97 \cdot 10^{-7}, 4.97 \cdot 10^{-10})$, resulting in an RSS of 0.011, a MAPE of 0.39 percent and a performance simulation shown in Figure 10.4.

Exercise

Table 10.3 provides an exemplary data set with 10 training loads (input), four performance measurements (output) and a parameter set for the IR Model. Use equations 3 to 7 to recursively determine simulated performance for days 2 to 10, starting with performance $p_0 = 40$ on day 1. Compute the RSS and the MAPE between empirical and simulated performance.

Table 10.3 Exemplary data and parameter set to use a performance model by hand.

Day	1	2	3	4	5	6	7	8	9	10
Training load (input)	5.5	8	7	0	5	2	0	8	7	0
Performance (output)	40	-	38	-	-	44.5	-	-	-	43
Simulated performance	40	?	?	?	?	?	?	?	?	?
Parameter	τ_1	τ_2	k_1	k_2	RSS			MAPE		
Value	45	10	0.2	0.1	?			?		

Notes

1 The synonym "training effect model" is rarely used and, therefore, omitted in the following.
2 Denoted in the literature by time constants or delays.
3 Alternatively denoted by scale, weighting, proportionality or multiplying factors, gain term or amplitude coefficients.
4 Also "Banister Model," "Fitness-Fatigue Model" (FF Model) and "Dose-Response Model."
5 Also "Variable Dose-Response Model."
6 Also referred to as goodness-of-fit.

References

Agostinho, M., Philippe, A., Marcolino, G., Pereira, E., Busso, T., Candau, R. & Franchini, E. (2015). Perceived training intensity and performance changes quantification in judo. *The Journal of Strength & Conditioning Research*, 29(6), 1570–1577.

Avalos, M., Hellard, P. & Chatard, J-C. (2003). Modeling the training-performance relationship using a mixed model in elite swimmers. *Medicine and Science in Sports and Exercise*, 35(5), 838.

Banister, E., Calvert, T., Savage, M. & Bach, T. (1975). A systems model of training for athletic performance. *Aust J Sports Med*, 7(3), 57–61.

Banister, E., Carter, J. & Zarkadas, P. (1999). Training theory and taper: Validation in triathlon athletes. *European Journal of Applied Physiology and Occupational Physiology*, 79(2), 182–191.

Banister, E. & Fitz-Clarke, J. (1993). Plasticity of response to equal quantities of endurance training separated by non-training in humans. *Journal of Thermal Biology*, 18(5–6), 587–597.

Banister, E. & Hamilton, C. (1985). Variations in iron status with fatigue modelled from training in female distance runners. *European Journal of Applied Physiology and Occupational Physiology*, 54(1), 16–23.

Banister, E., Morton, R. & Fitz-Clarke, J. (1992). Dose/response effects of exercise modeled from training: Physical and biochemical measures. *The Annals of Physiological Anthropology*, 11(3), 345–356.

Borresen, J. & Lambert, M. (2009). The quantification of training load, the training response and the effect on performance. *Sports Medicine (Auckland, N.Z.)*, 39(9), 779–795.

Busso, T., Häkkinen, K., Pakarinen, A., Carasso, C., Lacour, J., Komi, P. & Kauhanen, H. (1990). A systems model of training responses and its relationship to hormonal responses in elite weight-lifters. *European Journal of Applied Physiology and Occupational Physiology*, 61(1), 48–54.

Busso, T., Häkkinen, K., Pakarinen, A., Kauhanen, H., Komi, P. & Lacour, J. (1992). Hormonal adaptations and modelled responses in elite weightlifters during 6 weeks of training. *European Journal of Applied Physiology and Occupational Physiology*, 64(4), 381–386.

Busso, T. (2003). Variable dose-response relationship between exercise training and performance. *Medicine and Science in Sports and Exercise*, 35(7), 1188–1195.

Busso, T., Carasso, C. & Lacour, J-R. (1991). Adequacy of a systems structure in the modeling of training effects on performance. *Journal of Applied Physiology*, 71(5), 2044–2049.

Busso, T., Denis, C., Bonnefoy, R., Geyssant, A. & Lacour, J-R. (1997). Modeling of adaptations to physical training by using a recursive least squares algorithm. *Journal of Applied Physiology*, 82(5), 1685–1693.

Calvert, T., Banister, E., Savage, M. & Bach, T. (1976). A systems model of the effects of training on physical performance. *IEEE Transactions on Systems, Man, and Cybernetics, SMC*, 6(2), 94–102.

Candau, R., Busso, T. & Lacour, J. (1992). Effects of training on iron status in cross-country skiers. *European Journal of Applied Physiology and Occupational Physiology*, 64(6), 497–502.

Chalencon, S., Busso, T., Lacour, J-R., Garet, M., Pichot, V., Connes, P., … Barthélémy, J. (2012). A model for the training effects in swimming demonstrates a strong relationship between parasympathetic activity, performance and index of fatigue. *PLoS One*, 7(12), e52636.

Clarke, D. & Skiba, P. (2013). Rationale and resources for teaching the mathematical modeling of athletic training and performance. *Advances in Physiology Education*, 37(2), 134–152.

Collette, R., Pfeiffer, M. & Kellmann, M. (2016). *Trainings-und Erholungsmonitoring im Leistungssport Schwimmen*. Hamburg: Verlag Dr. Kovač.

Edelmann-Nusser, J., Hohmann, A. & Henneberg, B. (2002). Modeling and prediction of competitive performance in swimming upon neural networks. *European Journal of Sport Science*, 2(2), 1–10.

Ferger, K. (2010). Dynamik individueller Anpassungsprozesse: Eine vergleichende Analyse statistischer Zeitreihenverfahren und modellbasierter Dynamikanalysen. *Sportwissenschaft*, 40(1), 9–18.

Fitz-Clarke, J., Morton, R. & Banister, E. (1991). Optimizing athletic performance by influence curves. *Journal of Applied Physiology*, 71(3), 1151–1158.

Ganter, N., Witte, K. & Edelmann-Nusser, J. (2006). Performance prediction in cycling using antagonistic models. *International Journal of Computer Science in Sport*, 5(2), 56–59.

Haar, B. (2011). Analyse und Prognose von Trainingswirkungen: Multivariate Zeitreihenanalyse mit künstlichen neuronalen Netzen. Doctoral thesis, University of Stuttgart, Institute of Sports Science.

Halson, S. (2014). Monitoring training load to understand fatigue in athletes. *Sports Medicine (Auckland, N.Z.)*, 44 Suppl 2, S139–147.

Hausswirth, C. & Mujika, I. (2013). *Recovery for performance in sport*. Champaign, IL: Human Kinetics.

Hellard, P., Avalos, M., Lacoste, L., Barale, F., Chatard, J-C. & Millet, G. (2006). Assessing the limitations of the Banister model in monitoring training. *Journal of Sports Sciences*, 24(05), 509–520.

Hellard, P., Avalos, M., Millet, G., Lacoste, L., Barale, F. & Chatard, J-C. (2005). Modeling the residual effects and threshold saturation of training: A case study of Olympic swimmers. *Journal of Strength and Conditioning Research*, 19(1), 67.

Hohmann, A., Edelmann-Nusser, J. & Henneberg, B. (2000). A nonlinear approach to the analysis and modeling of training and adaptation in swimming. *Application of Biomechanical Study in Swimming.* Proceedings of The XVIII International Symposium on Biomechanics in Sports (pp. 31–38). Hong Kong: Chinese University Press.

Jobson, S., Passfield, L., Atkinson, G., Barton, G. & Scarf, P. (2009). The analysis and utilization of cycling training data. *Sports Medicine (Auckland, N.Z.), 39*(10), 833–844.

Kolossa, D., Bin Azhar, M., Rasche, C., Endler, S., Hanakam, F., Ferrauti, A. & Pfeiffer, M. (2017). Performance estimation using the fitness-fatigue model with Kalman Filter feedback. *International Journal of Computer Science in Sport, 16*(2), 42.

Kraemer, W. & Fleck, S. (2007). *Optimizing strength training: Designing nonlinear periodization workouts.* Champaign, IL: Human Kinetics.

Matabuena, M. & Rodriguez, R. (2016). A new approach to predict changes in physical condition: A new extension of the classical Banister model. *ArXiv Preprint ArXiv:1612.08591.*

McArdle, W., Katch, F. & Katch, V. (2010). *Exercise physiology: Nutrition, energy, and human performance.* Philadelphia, PA: Lippincott Williams & Wilkins.

Mester, J. & Perl, J. (2000). Grenzen der anpassungs- und leistungsfähigkeit des menschen aus systemischer Sicht. *Leistungssport, 30*(1), 43–51.

Millet, G., Groslambert, A., Barbier, B., Rouillon, J. & Candau, R. (2005). Modelling the relationships between training, anxiety, and fatigue in elite athletes. *International Journal of Sports Medicine, 26*(06), 492–498.

Millet, G., Candau, R., Barbier, B., Busso, T., Rouillon, J-D. & Chatard, J-C. (2002). Modelling the transfers of training effects on performance in elite triathletes. *International Journal of Sports Medicine, 23*(01), 55–63.

Morton, R. (1997). Modelling training and overtraining. *Journal of Sports Sciences, 15*(3), 335–340.

Morton, R., Fitz-Clarke, J. & Banister, E. (1990). Modeling human performance in running. *Journal of Applied Physiology, 69*(3), 1171–1177.

Morton, R. (1991). The quantitative periodization of athletic training: A model study. *Research in Sports Medicine: An International Journal, 3*(1), 19–28.

Mujika, I., Busso, T., Lacoste, L., Barale, F., Geyssant, A. & Chatard, J-C. (1996). Modeled responses to training and taper in competitive swimmers. *Medicine and Science in Sports and Exercise, 28*(2), 251–258.

Osterburg, A., Rojas, S., Strüder, H. & Mester, J. (2002). Adaptation research: Time-series-analyses in triathlon. A case study. In M. Koskolou, N. Geladas & V. Klissouras (Eds.), *Book of abstracts: 7th annual congress of the European college of sport science (ECSS)* (p. 693). Athens: ECSS.

Osterburg, A., Rojas, S., Strüder, H. & Mester, J. (2003). Neuroendocrine response to high training load on top performance level: A single case time series analysis. In E. Müller, H. Schwameder, G. Zallinger & V. Fastenbauer (Eds.), *Book of abstracts: 8th annual congress of the European college of sport science (ECSS)* (p. 149). Salzburg: ECSS.

Perl, J. (2001). PerPot: A metamodel for simulation of load performance interaction. *European Journal of Sport Science, 1*(2), 1–13.

Perl, J. & Pfeiffer, M. (2011). PerPot DoMo: Antagonistic meta-model processing two concurrent load flows. *International Journal of Computer Science in Sport, 10*(2).

Pfeiffer, M. (2008). Modeling the relationship between training and performance: A comparison of two antagonistic concepts. *International Journal of Computer Science in Sport, 7*(2), 13–32.

Pfeiffer, M. & Hohmann, A. (2012). Applications of neural networks in training science. *Human Movement Science, 31*(2), 344–359.

Philippe, A., Py, G., Favier, F., Sanchez, A., Bonnieu, A., Busso, T. & Candau, R. (2015). Modeling the responses to resistance training in an animal experiment study. *BioMed Research International 2015*, 914860.

Schwellnus, M., Soligard, T., Alonso, J-M., Bahr, R., Clarsen, B., Dijkstra, H., ... Hutchinson, M. (2016). How much is too much?(Part 2) International Olympic Committee consensus statement on load in sport and risk of illness. *Br J Sports Med*, *50*(17), 1043–1052.

Soligard, T., Schwellnus, M., Alonso, J.-M., Bahr, R., Clarsen, B., Dijkstra, H., ... Hutchinson, M. (2016). How much is too much? (Part 1) International Olympic Committee consensus statement on load in sport and risk of injury. *Br J Sports Med*, *50*(17), 1030–1041.

Taha, T. & Thomas, S. (2003). Systems modelling of the relationship between training and performance. *Sports Medicine*, *33*(14), 1061–1073.

Wallace, L., Slattery, K. & Coutts, A. (2014). A comparison of methods for quantifying training load: Relationships between modelled and actual training responses. *European Journal of Applied Physiology*, *114*(1), 11–20.

Part V
Tools and equipment used in sport

11 Modelling and simulation to prevent overloads in snowboarding

Veit Senner, Stefan Lehner, Frank I. Michel and Othmar Brügger[1]

Introduction: Models in the context of this chapter

Many publications and textbooks have given definitions of the term "model" and the vast variety of the given explanations shows that there is no common understanding of its definition. Instead, a context-specific explanation seems to be helpful. In this article, we will illustrate how different types of models have been combined in order to answer the given research question.

Our chosen path includes "close-to-reality experiments" with humans, which deliver important boundary conditions for the later mathematical representation of the human system of muscles, bones and soft tissue. Further, we combine two rather different mathematical approaches—multibody—and finite element modelling. While the first follows a more global view based on classical equations of motion (derived from either Newton-Euler or Lagrange's formalism), the second approach belongs to numerical mechanics using infinitesimally small elements and formulating conditions of equilibrium such as the sum of virtual work equals zero.

At this point, we would like to make the reader aware that, besides biomechanical models as shown in this book chapter, a multitude of other models and model categories from various research fields are required to entirely explain the interaction between humans and their equipment and the environment. Besides the contributions from biomechanics, we need models (developed and validated by psychologists) to explain human behavior; models (established in the field of kinesiology) that describe the neuromuscular control of human motion; and models (proposed by physiologists and biologists) that reproduce metabolism and energy transfer. Even though great advancements have been achieved within each of these research areas, we still have some way to go to bring them together into a holistic model of human motion.

A small step towards this goal was taken in the professorial dissertation of Boehm (2017) who described a set of different models including a perception model with the target to better understand and quantify functionality of sport equipment. Another interesting combination of different model categories has been developed by Fiala et al. (2010). Their research aims to simulate the thermo-physiology of the exercising or working human, considering the different ways of heat production and heat transfer phenomenon, taking into account the most important boundary conditions such as isolation of the clothing and the surrounding climate conditions.

Against this background, the following example of a biomechanical model approach is just a small piece in the puzzle of understanding the human in sports and exercise.

Project/problem: Multibody and finite element model of wristguards in snowboarding

Snowboarding is one of the most popular winter sports worldwide, particularly among adolescents and young adults. In the United States during the 2012/2013 season, approximately 7.3 million people practiced snowboarding, compared to 8.2 million skiers (Wijdicks et al., 2014).

Among all snowboarding injuries, the upper extremities comprise the body region most frequently reported injured. The risk of wrist injuries is especially high in snowboarding and much higher than in alpine skiing (Kim, Endres & Johnson, 2012).

The prevention of wrist injuries requires effective protection, which can be achieved by attenuating the peak force transmitted to the radius and ulna. Michel et al. (2013) found that several studies recommended wearing wrist protectors to minimize the incidence and severity of wrist injuries.

Wrist protectors of very different designs and materials have entered the market but, thus far, no generally agreed label has been introduced that would tell the customer the level of protection these products provide. In order to develop international standards for testing and classifying this type of protective gear, a better understanding of the interaction of different functional elements of wrist protectors and their potential to reduce loads beyond critical values is needed.

As a multitude of factors need to be considered and experiments exposing subjects to injury-critical loads being out of the question for ethical reasons, modelling and simulation is the method of choice. In the following section, we will go into more detail to develop the above outlined problem into a concrete research question.

Boundary conditions to detail the research question

Virtually all (96%) wrist injuries are induced by a fall (Russell, Hagel & Francescutti, 2007; Idzikowski, Janes & Abbott, 2000) and this may be one reason why experienced athletes tend to have fewer wrist injuries than beginners: 72% of all wrist injuries occur within the first 7 days of learning to snowboard (Sasaki, Takagi, Ida, Yamakawa & Ogino, 1999; Hagel, Goulet, Platt & Pless, 2004; Burkhart & Andrews, 2010).

Based on a literature review, backward falls result in twice as many fractures as forward falls (Deady & Salonen, 2010). In one of our former studies, we identified the backward fall with outstretched upper extremity joints being the worst-case scenario (Lehner et al., 2014). The peak force was calculated to occur

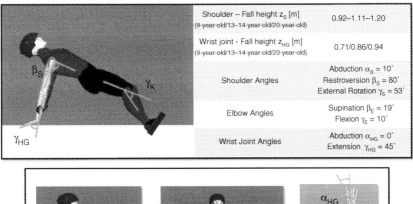
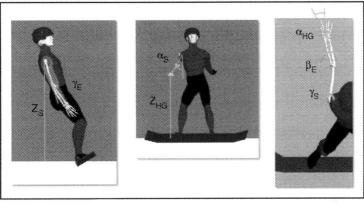

Figure 11.1 Backward fall starting condition: joint angles and joint-center location.

during hand impact with an outstretched elbow joint and an 80° retroversion of the shoulder joint. To get a better impression of the situation, the impact starting condition is illustrated in Figure 11.1. The angles defining this posture have not been chosen arbitrarily but from a video study with volunteers performing backward falls in the lab (Schmitt et al., 2012).

Bone strength changes with age and gender in children and adolescents. The major change is in the epiphyseal plate. In adults, when growth has stopped, the plate is replaced by an epiphyseal line.

Due to the enduring consequences of damage to the growth plates and the high incidence rate, the use of wrist protection devices is strongly recommended for children and adolescents (Michel et al., 2013).

The major functions of protector elements in snowboarding are to reduce impact force, to guide movements to avoid terminal extension and to prevent certain radial and pronation movements of the wrist and forearm.

There are many forms of wrist protection device currently available. It is important to know which types of device provide the necessary protection for this application.

A general distinction can be drawn between two principal design concepts: the "integrated protection concept," where protective elements are integrated into a glove; and the "separated protection concept" where protective elements are components of a separate wristguard (Schmitt et al., 2012).

To achieve a significant reduction in injury risk, a protector should provide sufficient impact attenuation and resistance to (hyper-)extension (Kroncke, Niedfeldt & Young, 2008). Impact forces can be reduced by energy absorption, energy redistribution or both.

With these statements in mind the research question we want to answer by means of modelling and simulation can be formulated as follows:

> *Which type of design or which functional elements of snowboard wrist protectors can reduce stress in the wrist and forearm bones in a typical backward fall situation, in particular for the target group children and adolescents?*

Concepts and methods of modelling

The above research question bares four major challenges which determine the basic concept of our model approach. First, we have to reproduce the kinematics and dynamics of the backward fall. Besides knowing the typical initial conditions of such a fall, this task also requires including target group-specific anthropometry (i.e., segment length, masses and inertia). These different segment parameters result in a different location of the center of mass and this means different initial potential energy. Second, we need to model the different functional elements of the wrist protector. This involves replicating the exact geometry of each component (i.e., padding) and describing their mechanical properties (stiffness and damping parameters). The third challenge is a realistic connection and fixation between the protection gear and the arm. To fulfill this requirement, the arm needs a surface with correct shape and volume and, equally important, we have to think about the degrees of freedom and relative displacements the wristguard might have against the arm. It is clear that a close-to-real description of the connection is essential for simulating the correct load pattern transmitted to the segment. This defines the questions to solve in the next challenging step: How will these loads be distributed to the bones and which strains and stresses will come of that? As we have stated in the beginning, this has to be done taking into account the very different bone structure and bone properties of children, adolescents and young adults. The age definition of these three groups, the rationale behind these and the chosen anthropometry are given in Table 11.1.

In order to meet the above four challenges we have chosen to combine different Computer Aided Engineering (CAE) tools as illustrated in Figure 11.2:

- Volume models of wrist protectors and of all human segments were developed using computer aided design (CAD) software, CATIA V5 (Dassault Systèmes, Vélizy-Villacoublay, France).
- Detailed 3D models were constructed of the upper extremity. The exact surface models of the arm, hand and finger bones as well as of the solid portions of the radius and ulna were created based on computer tomography (CT)

Table 11.1 Anthropometric data for the modeled age groups from the Size Germany database. The data represent the 50th percentile for males.

Age	Children	Adolescents	Young Adults
	9 years old	13–14 years old	20 years old
Rationale for age selection	Recommended best starting age for snowboarding, recommended by the German Ski Association (DSV). Epiphyseal plate present	Within this age range, boys increase height and mass at peak rates—2 years later than for girls.	At this age, the epiphyseal plates are closed and replaced by epiphyseal line.
Height (m)	1.39	1.67	1.80
Weight (kg)	32	56	75

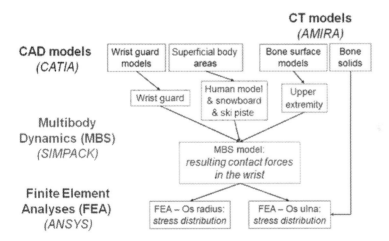

Figure 11.2 Overview of the chosen modelling concept. Different types of computer-aided engineering tools were used to model and simulate the potential injury situation and to calculate the resulting stress in the bones depending on the type of wrist-protection devices used.

data. Image segmentation and geometry reconstruction was done using the software system AMIRA (Thermo Fisher Scientific, Waltham, MA, USA).

• Fall simulations and calculation of all bone loads were conducted with the multibody body simulation (MBS) software, SIMPACK 9.3.2 (Dassault Systèmes, Vélizy-Villacoublay, France).

• A detailed analysis of the strain and stress on the forearm bones was conducted with finite element analysis (FEA) models (ANSYS Workbench 14.5, Ansys Inc., Canonsburg, USA). We used FEA models of the Os radius and Os ulna.

Figure 11.3 Backward falls simulated with MBS models, based on age-related anthropometries: (left) child – 9 years old; (middle) adolescent – 13 years old; (right) adult – 20 years old.

Today's MBS software packages, like the one we have been using, allows to parametrize such kinds of input data. This means that when building up the model on the computer we first define variables for all segment parameters and then give them certain values (that are clearly represented in a list at the beginning of the program code). This helps to easily change the model without going through the entire code, thus being able to quickly run simulations for different anthropometry.

Besides the human body with segments of different model complexity, our MBS also includes a surface model of the ski slope and simple model of a snowboard. More details will be given below.

Figure 11.3 illustrates the graphical representation of the initial conditions for the simulated backward falls for our three examined age groups.

Some details on the human segment model

The parametric human model consists of 12 rigid bodies, each body with realistic mass and inertia distribution. One main principle when developing models is to begin with simple models and increase their complexity step by step. One reason to follow this principle is given by the inherent process of error analysis: With complex models consisting of many different sub-models, it is much more difficult to analyze relationships between model parameters, chosen boundary conditions and the model's output. With less complex models, it is also easier to interpret simulation results and to verify correctness, robustness and principal model behavior. This recommendation however also includes the demand of going into detail if the given research question needs it, e.g., to achieve a certain accuracy or sensitivity. In our case, we have to go into detail when it comes to the anatomy of the arm, but we do not have to realize a sophisticated model of the leg. In order to follow the principle of simplicity (which also reduces computer calculation time) we have decided to implement the detailed arm model only for

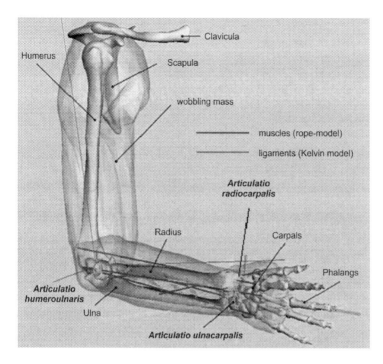

Figure 11.4 Detailed MBS model of the right upper extremity.

the right half of the body. This corresponds to another principle when modelling and simulating injury situations: Always start to model and simulate worst-case scenarios. In our case, the worst case is the backward fall on one extended arm and not transferring energy by the other.

The ranges of motion and joint stiffness are described with torque elements. The detailed MBS model of the right upper extremity includes surface models of the Os humerus, Os radius and Os ulna. The hand and finger bones were modeled as rigid bodies (Figure 11.4).

The connections between the soft tissues and the bones in the upper extremity were modeled as wobbling masses via spring-damper elements (Gruber, Ruder, Denoth & Schneider, 1998). Implementing wobbling masses—the soft tissue's motion relative to the bone—is not always necessary. But in cases with high dynamics like in our impact situation, it is essential to account for the fact that the skeleton will be decelerated in the first instant of time and then, with some time delay, the soft tissue follows. As a consequence, the calculated time history of the joint loads is more realistic and the maximum joint loads are significantly lower if we model wobbling masses. Generally, muscle contraction reduces wobbling of the soft tissue to a certain extent which, on the other hand, would increase impact loads. At the same time, this muscle contraction helps to stabilize joints and thus, in part or totally, compensates the aforementioned effect. In our case, the worst-case scenario is looking at the situation without fast muscle reaction. This assumption is rather common in accident and injury research either because the event is too short to trigger muscle reflexes or because the person is frightened and not able to adequately react.

In our model, the kinematic chain in the right upper extremity included the shoulder, with 3 rotational degrees of freedom (DoF); the elbow, with 1 rotational DoF; the forearm, with 1 translational DoF between the radius and ulna and with 1 pronation/supination DoF; the wrist with 2 rotational DoF; the distraction of the proximal row of the carpal bones on the radius; and finger joint with 1 rotational DoF. The membrane interossea antebrachii and the essential ligament structures in the wrist were modeled with visco-elastic elements. These elements' development and their important validation against the force-elongations measured in *in vitro* experiments with human or animal ligament samples is described in detail by Lehner (2008).

The contact surfaces in the wrist joint (i.e., in the *Articulatio radiocarpalis* and the *Articulatio ulnocarpalis*) and those in the elbow (i.e., in the *Articulatio humeroulnaris*) are modeled as areas of contact. Several potential contact points are included in the joints with unilateral elements, including a spring and a damper in parallel. Based on experimental data, these validated cartilage models required a stiffness of c = 30,000 N/m and a damping factor of d = 60,000 N/ms (Yamada, 1970).

Model for hand contact with the slope surface

The contact between the hand and the slope surface was based on a model published by DeGoede and Ashton-Miller (2003). The resulting contact force was expressed as follows:

$$F_{BRK} = K_{hand} \left|x_{hand}^3\right| (1 - B_{hand} \dot{x}_{hand})$$

where x_{hand} is the vertical displacement of the wrist joint relative to the initial position. The following parameters were used to model an impact on a hard surface:

$$K_{hand} = 10,000 \frac{kN}{m^3}; B_{hand} = 7\left(\frac{m}{s}\right)^{-1}$$

Wrist protector models

The current market of wrist protectors sees a vast variety of different concepts and models. To reflect this diversity, we developed a total of 20 models of different principal concepts described in Table 11.2 and illustrated in Figure 11.5a and 11.5b.

Preventing overloads in snowboarding 219

Table 11.2 Overview of modeled wristguard concepts and versions.

#		Version		
		short	medium	long
0	No wristguard (used as reference)	-	-	-
1–3	Wristguards with dorsal splints	✓	✓	✓
4–6	Wristguards with palmar splints	✓	✓	✓
7, 8	Sandwich configuration wristguards, with dorsal and palmar splints	-	✓	✓
9, 10	Wristguards with dorsal splints and 5 mm palm padding	-	✓	✓
11, 12	Wristguards with dorsal splints and 10 mm palm padding	-	✓	✓
13, 14	Wristguards with palmar splints and 5 mm palm padding	-	✓	✓
15, 16	Wristguards with palmar splints and 10 mm palm padding	-	✓	✓
17, 18	Sandwich configuration with 5 mm palm padding	-	✓	✓
19, 20	Sandwich configuration with 5 mm palm padding	-	✓	✓

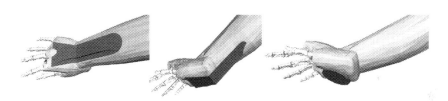

Figure 11.5a Wrist splint models. Splints are shown in gray: (left) dorsal splint, (middle) a palmar splint and (right) palm padding.

Figure 11.5b Fixation of dorsal splint to the forearm. Numbers and arrows indicate the positions of fixation: (left) the short version, (middle) the medium-sized version, and (right) the long version.

The protectors for the dorsal and palmar splints included two hinged joints, one at the top of the wrist joint and a second one in the fixed part of the forearm. The position of the second hinged joint depended on the splint length (short, medium or long). Within the hinged joints, a rotational spring-damper element was included to limit the range of motion. The joint material was described as thermoplastic polyester plastic Hytrel© 4053FG (DuPont). This chosen material is rather common. With the existing model, it is possible to simulate other materials (such as carbon-fiber composites offering substantially higher stiffness and providing the opportunity to design according to the stress situation present).

The dorsal and palmar splints were fixed to the forearm with a Velcro® fastener. This was modeled with four spring-damper elements, located on the circumference of the CAD surface of the forearm. The number of modeled hook-and-pile fasteners was varied; one was used for the short splints and two for the medium and long splints (Figure 11.5b).

The described fixation models reproduce the most important boundary conditions of the real world but do not reflect possible details such as small modifications in the position of the straps. Varying these values to analyze possible effects could be a logical next step for future applications of the model.

Using the above described fixation models, each of these wrist protectors were then implemented into our MBS model and the backward fall simulations were carried out for the three age groups (thus resulting in a total of 63 conditions).

Today's MBS software packages allow to define output parameters, for example the components of loads at a specific location. As we are interested in what happens at the wrist, we define as a crucial output parameter the resulting force (magnitude and direction) at the forearm bones. The time history of this parameter is then used as input for the following FEA. Details of this are explained in the following chapter.

A few details on the finite element model

The resulting force acting on the forearm bones are analyzed in more detail with FEA models of the Os radius bone. For the development of this FEA model and the consequent simulations, we use the software package ANSYS Workbench Version 14.5 distributed by Ansys Inc., Canonsburg, USA. Building up FEA models generally starts with implementing a realistic 3D-volume model of the body of interest, mostly given by a set of CAD data. In biomechanics, however, an additional step is required to obtain the required CAD data of structures like human bones. First, a complete set of sectional images of the entire arm using CT has to be gathered. It is clear that the quality of the later reconstruction highly depends on the step width during this imaging process. The step width depends on the desired accuracy of the volumetric reconstruction. In the zone of the growth plate where a more detailed reconstruction is needed, a step width in the magnitude of millimeters is recommended. In areas with minor changes in the geometry, i.e., the bone shaft, the step width might be reduced to centimeter level. However, it is possible to skip pictures (in order to reduce calculation time) and, often, we do not really know the geometry of inside structures (in our case, for instance, the changes in the thickness of the corticalis), so the general rule can be formulated: The smaller the step width, the better. In the second step, in every single CT image, both the outer and inner contours of the compact bone,

Table 11.3 Young's modulus for the different materials.

	Modulus values in [MPa] for age group		
	20 years old	13–14 years old	9 years old
Isotropic corticalis	18,000	15,000	12,000
Orthotropic corticalis x	20,000	15,000	12,000
Orthotropic corticalis y	13,000	9,750	5,850
Orthotropic corticalis z	11,000	8,250	4,950
Cortical-spongy transition	1,800	1,000	500
Cancellous bone	30	3	3
Epiphyseal plate	not applicable	15	12

as well as the dividing lines—for example with the cancellous bone (spongiosa)—have to be digitized. As this is a rather time-consuming task, we use the help of a special software package (AMIRA, FEI Visualization Sciences Group, Bordeaux, France), which allows to mostly automatize this process. In the last step, the digitized contours are then reconstructed to a 3D volume that can be used in common CAD programs.

Our FEA models include the behavior of the compact bone (corticalis) and of different trabecular structures. The chosen Young's moduli for the three different age groups are given in Table 11.3. Poisson's ratio and density of these materials have also been implemented using age-specific values. Furthermore, we distinguish the following four sections (Figure 11.6) of the radius bone:

- The shaft, which consists of orthotropic compact bone. The medulla is not included because its Young's modulus is very small thus having negligible effects;
- The proximal sector, which consists of a sector of isotropic compact bone and a sector of isotropic cancellous bone;
- The distal sector, which consists of isotropic compact bone, isotropic metaphysis defined as spongy material and isotropic epiphysis defined as a cortical-spongy transition;
- An epiphyseal plate for the 13–14-year-old boy and for the 9-year-old boy; this was modeled as two hyaline cartilage plates for both metaphyses of the radius.

Apart from choosing the values for material properties, it is also necessary to decide which FEA principle method should be used. In our case, Leitwein (2015) has switched from the earlier quasi-static mechanical structure analysis method to an implicit method (transient analysis). This not only allowed to consider dynamic effects like damping and inertia but also bares the possibility to significantly reduce the mesh size at specific locations of the structure. The disadvantage of the transient method—considerably longer calculation time—had to be accepted.

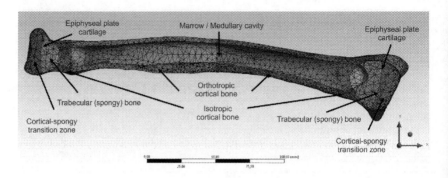

Figure 11.6 Zones and bone tissues distinguished in our FEA model of the radius. The picture shows the version used to investigate the situation for the group of 20-year-olds. For them, no epiphyseal plate and no marrow are modeled.

As expressed in the term "finite element modelling," it is necessary to represent the structure of interest by a closed 3D net of small volumes, called voxels. The process of filling the volume adequately with these kind of elements is often called "meshing." Neither the geometric type nor its dimensions are fixed *a priori*, but they are also far away from being arbitrary. The calculation time for the FEA increases exponentially if the mesh becomes finer. We thus have to find a good compromise between mesh resolution and calculation time. Several ways exist to systematically find this compromise (Gebhardt, 2014). Performing in the transient structure analysis mode, the used version of ANSYS Workbench does not allow to perform a classical convergence test (reducing the size of the voxels step by step until the differences between the calculated solutions converge below a defined threshold). We therefore had to use other criteria, such as the median of the voxels' aspect ratio distribution and the smoothness of the calculated stress distribution. As a result of this analysis, the bone was meshed with 4.8 mm tetrahedral elements and the area of the epiphyseal plate with 2.0 mm elements of the same type.

After having meshed the structure, we have to program which external loads are acting from which direction and with what magnitude. In our case, several forces are acting on the radius, due to soft tissue structures and due to energy transfer over joint contacts. Seven contact forces are generated by the carpal bones, 10 tension forces by the distal ligaments and five by the ligaments in the shaft area. Each point of force application is related to the anatomical region as given by anatomy books. The calculated forces that affect the radius are transferred from the MBS to the FEA model. It should be mentioned at this point that this procedure simplifies the real world: Instead of modelling surface-load distributions typical for ligament insertion areas, we represent each ligament by a single vector acting at a defined position (marker). In contrast to this, and as described above, the bone-to-bone contacts are modeled by a set of several contact elements.

To determine whether a fracture of the radius could occur in the simulations of backward falls, the calculated stresses were compared to the ultimate stresses for cortical bone, published by Reilly and Burstein (1974). Because shear stress is an invariant value that depends on physical properties only, we use shear stress as a relevant parameter to evaluate the behavior of the bone in the FEA.

Validation of our finite element model

The explanations given above on how to derive an FE model of a complex skeletal structure interacting with soft tissue should have clarified: a multitude of material parameters exist, each having a wide possible value range due to the typical high variation in humans; that necessary simplifications and inherent inaccuracies may in the sum lead to unrealistic simulation results. It is therefore a basic requirement to put much effort into the validation (verification) of a new model. The majority of our partial models (i.e., our ligament model) have been verified against experimental data. To limit the complexity of this chapter, details on the validation are given for the FE model only.

Our FE model of the 20-year-old was validated by remodelling and simulating compression tests with human cadaver bones, conducted and published previously by Varga, Baumbach, Pahr and Zysset (2009) and Zysset et al. (2015). In their investigation, a total of n = 21 embalmed human radii were submitted to an axial compression force. The maximum linear stiffness achieved in these experiments was 8.4 N/mm. In their FE approach, which described an experimental model of Colles' fractures, they calculated a maximum linear stiffness of 12.9 N/mm. With our FE model simulating comparable boundary conditions as used by Varga and Zysset, a maximum linear stiffness of 10.3 N/mm was calculated.

	Maximum linear stiffness [N/mm]
Experimental data – Varga et al. (2009)	8.4
FE model – Varga et al. (2009)	12.9
FE model – **our study** (for 20-year-old)	**10.3**

In interpreting such kinds of comparisons between experiments and FEmodelling and subsequent simulations we have to be aware of the fact that not just the model, but also the experiment is only able to provide an approximation of the reality: Experimental tests with human or animal cadaveric specimens are subject to well-known limitations such as disease transmission, high variability with respect to their mechanical properties, decay and fatigue behavior for longer testing procedures (Hamer et al., 1996; Perrott, Richard, Smith, Leonard & Kaban, 1992; Woo, Orlando, Camp & Akeson, 1986).

It is therefore consequent that literature gives wide ranges for characteristic values like the ultimate stress for human cortical bone. Reilly and Burstein's (1974) fundamental work, for example, ranges the ultimate pressure for cortical bone from 106 MPa to 196 MPa.) Keeping these typical magnitudes of variation for biomechanical failure tests in mind, the difference of 20% between our simulation and Varga's experiment seems acceptable.

Unfortunately, no experimental data is available in literature for validating our FE models with the epiphyseal plates for the 13–14-year-old and the 9-year-old boys. This is a clear limitation in our approach, however there is no solution available at the moment.

Results

Multibody dynamics

The backward fall was simulated with 20 models of wrist protectors based on different principal concepts, including dorsal and/or palmar splints, different splint lengths (short, medium, long), and without or with palm padding. As a reference, we use the parametric human MBS models of the three boys of different age groups falling without wrist protection.

The most common injury mechanism of a radial fracture is described as a comprehensive load to a hyperextended wrist (Michel et al., 2013). Table 11.4 shows the calculated maximum compressive (axial) loads and the maximum wrist dorsiflexion (also named "wrist extension") at impact.

Figure 11.7a shows the relative reductions in wrist extension in case protectors are used (not considering the additional influence of padding) compared to the reference models without protectors. The short splints—both palmar and dorsal—reduce the wrist extension in all age groups by a maximum of 11.3%; the long versions reduce wrist extension between 18.6% and 25.4%. All simulations with the adult model (20-year-old) show that any long or medium splint variation (dorsal, palmar, sandwich) provides a reduction in dorsiflexion of at least 18.4%. Among the long protectors, the palmar version reduces the wrist extension less than the dorsal and sandwich versions. No substantial differences are observed between the dorsal and the sandwich versions for all three age groups.

Padding affects all models in all three age groups but the efficiency of padding to limit wrist extension is rather small: 2.1% to 2.5% relative reductions in wrist extension for the adult and adolescent models and only 0.7% to 1.3% relative reductions for the child model. No difference in this trend is observed between long and medium versions or between dorsal and palmar variations. In contrast to its minor effects to limit wrist extension, padding seems to be more efficient to reduce the axial force at the wrist (see below).

Without a protective splint, upon impact, the calculated compressive (axial) load applied to a hyperextended wrist is 2547 N in the adult model (20-year-old), 1780 N in the adolescent model (13–14-year-old) and 981 N in the child model (9-year-old) These axial loads calculated for the unprotected models are used as references.

Figure 11.7b illustrates noticeable relative reductions in maximum wrist joint force (simulated without additional padding) for both adults and adolescent male

Table 11.4 Maximum load at the wrist joint and maximum wrist extension during impact.

Age group	Axial loading (N)			Wrist extension (deg)		
	20 years	13–14 years	9 years	20 years	13–14 years	9 years
No wristguard						
Reference	2,547	1,780	981	69.9	68.5	64.7
Dorsal splints						
Dorsal short	1,413	1,194	951	62.0	62.1	61.4
Dorsal medium	1,407	1,092	667	55.4	54.7	54.2
Dorsal long	1,213	990	616	52.8	52.0	51.0
Palmar splints						
Palmar short	1,367	1,055	827	63.9	63.5	62.0
Palmar medium	1,465	1,106	678	57.1	56.0	56.1
Palmar long	1,379	1,071	653	54.2	53.0	52.6
Sandwich (palmar & dorsal splints)						
Sandwich medium	1,395	1,117	670	55.1	53.9	54.1
Sandwich long	1,189	1,031	607	52.1	51.6	50.8
With padding – 5 mm thick (with medium splints)						
Dorsal medium & padding 5 mm	932	868	711	53.9	53.2	53.8
Palmar medium & padding 5 mm	999	817	749	55.5	54.5	55.4
Sandwich medium & padding 5 mm	952	916	707	53.7	52.5	53.7
With padding – 10 mm thick (with medium splints)						
Dorsal medium & padding 10 mm	952	869	691	53.8	53	53.6
Palmar medium & padding 10 mm	1,023	809	736	55.3	54.5	55.3
Sandwich medium & padding 10 mm	955	916	686	53.5	52.3	53.5
With padding – 5 mm thick (with long splints)						
Dorsal long & padding 5 mm	941	962	620	51.6	50.8	50.8
Palmar long & padding 5 mm	1,011	979	723	52.7	51.7	52.3
Sandwich long & padding 5 mm	1,043	1,031	623	51	50.5	50.6
With padding – 10 mm thick (with long splints)						
Dorsal long & padding 10 mm	941	962	624	51.4	50.6	50.7
Palmar long & padding 10 mm	1,013	979	707	52.6	51.5	52.2
Sandwich long & padding 10 mm	1,043	1,031	623	50.9	50.3	50.5

and for all protector versions simulated. The maximum effect is observed for the adult model with reductions greater than 50%. For the child model, however, the short versions of the protectors show substantially smaller effects than for the medium and long versions.

Padding may provide additional reductions to the wrist loads, however less pronounced for the children's versions than for the adolescents and adults. For the adults' medium dorsal splint version, for example, 5 mm padding reduces the peak force at the wrist by an additional 18% (1407 N without padding; 932 N with 5 mm padding) and by an additional 12% (1092 N w/o; 868 N with

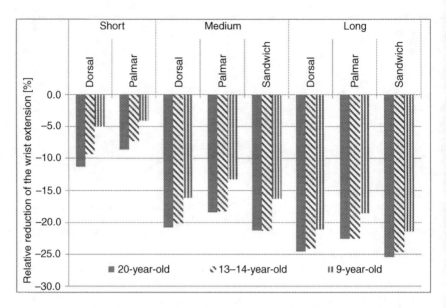

Figure 11.7a Relative reductions (%) in wrist extension when different protector designs are compared to the reference model without a protector (0.0). All protectors are versions without padding.

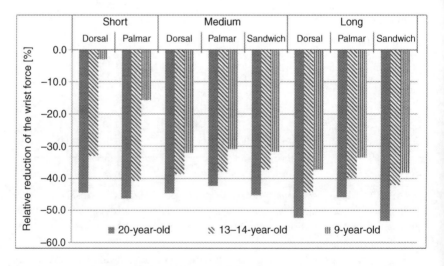

Figure 11.7b Relative reductions (%) in axial wrist joint force for different protector designs compared to the reference model without a protector (0.0). All protectors are simulated without padding.

padding) for the adolescent model. In contrast, the same protector with additional padding but applied to the children's model even slightly increased the axial load by 4% (667 N w/o; 711 N with padding). For the palmar protector version, the additional relative reductions are 17% for the adult model, about 16% for the adolescent model but less than 7% for the child model. No significant differences are observed between the 5 mm and 10 mm padding versions.

Keeping young beginners in mind, the most important result is that the long dorsal and sandwich versions offer the most effect, reducing both the axial load and the resulting dorsiflexion. In clear contrast to this, the short versions seem to be rather ineffective. For adults and adolescents, the long dorsal splint and the long sandwich version seem to be best with regard to limiting both axial force end wrist extension. For adults, the most pronounced load reduction is given by either the 5 mm padded medium-sized or the 5 mm padded long dorsal splint version (both 63% less than the reference).

Finite element analyses

Based on the results of the MBS simulations, the FEAs were carried out with all age models. The validated adult FE model (see *Validation of our finite element model*) was used to analyze all versions (short, medium, long) of the palmar and dorsal splints. The child and adolescent models were only used to analyze the long splints, which showed the most promising effects when simulating the adult model.

Table 11.5 shows the maximum stresses calculated for the adult model in the reference condition (with no protector) and when wearing each of the three versions (short, medium and long) of the dorsal and palmar protectors. The stresses calculated for the reference model were at the lower boundary of the ultimate shear stress for cortical bone, published by Reilly and Burstein (1974).

Simulating the reference condition, the validated adult model calculates a maximum shear stress of 62.7 MPa at the palmar radius, located just above the joint space at the wrist. The maximum stress is observed in the inner layer of the compact bone (Figure 11.8a). If the adult model "wears" a long dorsal protector, the maximum shear stress decreases to 43.7 MPa. The location of maximum stress

Table 11.5 Maximum shear stresses calculated in the FEAs with the validated adult model (20 years old) in the reference condition (with no protector) and when wearing each of the three versions of the dorsal and palmar protector models.

	Ultimate stress	Reference (without)	Dorsal			Palmar		
			short	medium	long	short	medium	long
Shear stress (MPa)	53–82	62.7	60.5	61.5	43.7	60.5	59.2	53.4

Note: For comparison, the range for ultimate stress of cortical bone is given, based on the published values by Reilly and Burstein

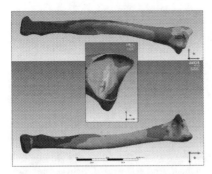
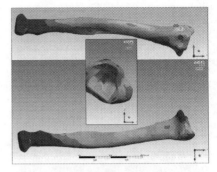
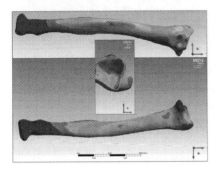
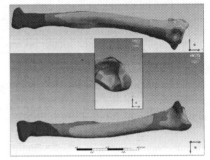

Figure 11.8 The adult group's FEAs of stress distribution in the radius bone. Each picture shows the dorsal view of the bone at the top and the palmar view at the bottom. In the middle, the zoomed radial corticalis is shown from the inside. Maximum shear stress appears at different locations: (a-top left) reference condition (no wristguard), (b-top right) short dorsal splint, (c-bottom left) medium dorsal splint and (d-bottom right) long dorsal splint.

remains in the inner layer of the compact bone, just above the joint space at the wrist. The maximum, however, moves slightly from the palmar radius (with no protector) to a location above the styloid process of the radius and an increased shear stress can be observed in the shaft (Figure 11.8d).

In case the adult model is using a long palmar protector, the maximum shear stress is calculated with 53.4 MPa. Similar to the dorsal splint, the palmar splint shifts the stress distribution. Again, the maximum stress is located above the styloid process of the radius and the shear stress is increased in the shaft.

As illustrated in Table 11.5, the short- and the medium-length versions of both protector types—palmar and dorsal splint—show only minor stress reduction effects. We calculate the short dorsal splint reducing the maximum shear stress from 62.7 MPa (reference condition) to 60.5 MPa only. The maximum stress is located above the styloid process of the radius and the shear stress is increased in the shaft just above the end of the splint (Figure 11.8b). The medium-length splint also results in only a minor reduction in maximum shear stress (61.5 MPa),

Table 11.6 Stress reduction (%) relative to the reference model for the adult, the adolescent and the boy wearing the long versions of the dorsal and palmar protectors.

Age group	Long dorsal protector			Long palmar protector		
	20 years old	13–14 years old	9 years old	20 years old	13–14 years old	9 years old
Change in maximum shear stress (%)	–30.3	–43.0	–54.2	–14.8	–52.0	–53.8

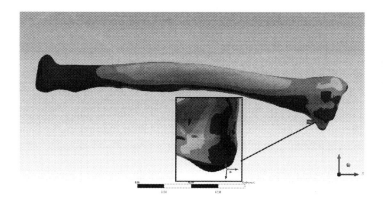

Figure 11.9 FEAs of the stress distribution at the radius in the adolescent model when wearing the long dorsal protector.

but the stress is increased in two areas on the shaft. Here, the absolute maximum has shifted to the shaft, above the end of the splint (Figure 11.8c).

For the 13–14-year-old adolescent and the 9-year-old boys, we compared the load distribution observed in the reference condition to those observed when wearing the long dorsal protector and the long palmar protector (Table 11.6). As mentioned before, experimental data are not available for validating these FE models. Therefore, a comparative study was carried out only, with relative reductions in stresses. For the adolescent model, the palmar version showed the higher stress reduction. For both long splint child models, stress reduction is greater than 50%, the dorsal and palmar models being not substantially different.

Looking at the locations of maximum stress distribution calculated for the adolescent model in the reference condition shows maximum stress in the hyaline cartilage plate region. The long dorsal protector reduces the maximum shear stress at the hyaline cartilage plate by 43%. Again, shear stress was increased in the shaft (Figure 11.9).

The stress distribution calculated for the 9-year-old boy in the reference condition also shows maximum stress at the hyaline cartilage plate. The long dorsal protector reduces the maximum shear stress at this location by 54% (Figure 11.10).

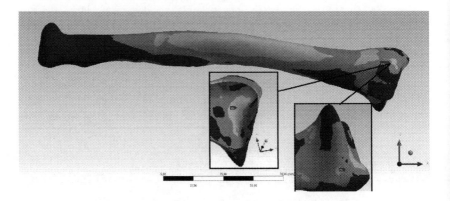

Figure 11.10 FEAs of the stress distribution at the radius in the child model when wearing the long dorsal protector.

Discussion

A new workgroup (CEN/TC 162/WG 11) has been established February 2016 to continue project ISO 20320 "Protective clothing for use in Snowboarding – Wrist Protectors – Requirements and test methods." In part, this wrist-guard standard compares to the existing standard EN 14120 for roller sports enthusiasts, which defines requirements and methods for testing protective clothing and wrist protectors. In the white paper for snowboarding (Michel et al., 2013), initial considerations are given with regard to possible standard requirements for the two major wrist protector performance variables "impact performance" and "bending stiffness." According to the EN 14120 standard, the performance criterion is fulfilled for a wristguard that limits movement within the 40–55° range, in response to an applied moment of 3 Nm (European Committee for Standardization 2007).

Greenwald et al. (1998) measured wrist extension and forces at the wrist joint in on-slope experiments with an instrumented snowboarding glove. In their study, adult volunteers (>17 years old) experienced significantly higher maximum force compared to adolescents (≤17 years). However, even though a maximum force of 600 N was recorded during falls on the slope, the wrist impact forces were much lower. In other laboratory experiments that tested different drop heights and different impact surfaces, the measured loads were between 700 N and 1000 N (Lewis, West, Standeven & Jarvis, 1997; Moore, Popovic, Daniel, Boyea & Polly, 1997; Chiu & Robinovitch, 1998; Staebler et al., 1999; DeGoede & Ashton-Miller, 2002; Kim, Alian, Morris & Lee, 2006; Schmitt, Michel & Staudigl, 2011). Greenwald, Janes, Swanson & McDonald (1998) recorded maximum extension moments (torques) of 10.8 ± 11.3 Nm for adolescents (≤17 years) and 16.5 ± 20.7 Nm for adults (>17 years). Recordings from seven adult snowboard beginners (mean body mass 72.9 kg) during backward falls showed maximum forces of 300.9 ± 284.5 N, maximum extension moments of 15.9 ± 20.7

Nm and maximum wrist extensions of 76.8 ± 15.8°. However, recordings from two adult advanced snowboarders (body masses 69 and 82 kg) during backward falls showed maximum forces of 695.0 ± 69.3 N, increased maximum extension moments (35.5 ± 0.7 Nm) and decreased maximum wrist extensions (51.5 ± 5°). No records showed load magnitudes sufficient to fracture the radius during these on-slope experiments, and no fractures were reported. However, an analysis of the fracture risk and a valid test for the performance of different wrist-protector models would require more realistic data, in the range of potential fractures.

With the MBS developed in the present study, a backward fall was evaluated as the worst-case scenario among different types of snowboarding falls (Lehner, Geyer, Michel, Schmitt & Senner, 2014). Based on our model of a 20-year-old male adult, we calculated that the main impact load was 2547 N in the wrist joint and the maximum extension moment was 52.5 Nm at impact. Loads of this magnitude bare the risk of fracturing the radius. Several laboratory studies that investigated wrist loads in simulated falls with either mechanical surrogates or cadaveric arms demonstrated that loads in the order of 2.0–2.5 kN were sufficient to fracture the radius.

Another criterion for evaluating wrist-guard performance is the bending stiffness. The bending stiffness is evaluated by the bending angle at the wrist joint. In the present comparative study, only one falling situation was simulated. The reference was the dorsiflexion calculated in a model with no protective gear. The reference dorsiflexions were 69.9° for the adult model (20-year-old), 68.5° for the adolescent model (13–14-year-old) and 64.7° for the child model (9-year-old). In general, the normal range of motion in the wrist joint is 60–75°; therefore, the dorsiflexion of 70° calculated in the adult model simulation during a backward fall was within the range limit for causing a hyperextension injury. The maximum possible wrist extension is considered to be 80°, given the wrist bending angle. Our finding of a 70° wrist extension during simulated falls was consistent with the range of maximum recorded wrist extensions (80 ± 15°) in on-slope experiments carried out by Greenwald, Michel and Simpson (2013) and in laboratory tests carried out by Schmitt et al. (2012) (left wrist: 85 ± 7°; right wrist: 82 ± 7°).

In the absence of padding, our bending results show that the short splint versions do not provide sufficient protection, but the protection effect increased with the longer versions. The dorsal splint and the sandwich splint show a slightly decreased wrist angle at impact compared to the palmar splint. This protective effect is enhanced with padding, but there is no significant difference between the 10 mm and the 5 mm padding thicknesses.

Our investigation was not designed to analyze the comfort of wearing the protectors. However, we found that due to the stiffness of the protector models, the end of the dorsal splints penetrates into the soft tissue (wobbling mass) of the forearm (Figure 11.11). The maximum penetration measured at impact is 6 mm for the short and long splint versions and 8 mm for the medium splint version.

As is the case for all studies, the present one has some limitations. One weakness of the FEA model is that it does not accommodate the fact that bone

Figure 11.11 Displacement of the dorsal protector at impact. The proximal end of the splint penetrates into the wobbling mass of the forearm.

is a non-homogeneous, anisotropic material with mechanical properties that change as a function of the specific location of the bone and the direction of force application. Our assumption of linear isotropic material behavior and homogeneous material distribution within each of the distinguished zones is a decisive simplification of the real world and especially neglects visco-elastic properties of the bone tissues. To address this simplification, we have chosen values for Young's modulus from the upper end of the given ranges in literature. To some extent, this takes into account the stiffening effect of increased strain rates as described by Wright and Hayes (1976).

In addition, due to the lack of data on fracture loads of the radius for adolescents and children, we could not validate FE models that included the epiphyseal plate for the 13–14-year-old and the 9-year-old boys. Nevertheless, the locations of the maximum stresses calculated in this comparative FE study appear to be realistic. For the validated adult model, in the reference condition, the maximum shear stress occurred at the palmar radius, just above the joint space. The maximum is observed at the inner layer of the compact bone in the model. The calculated stress distributions for the reference adolescent model and for the reference model of the child show maximum stress at the hyaline cartilage plate.

Conclusion

CAE tools enabled us to evaluate different types of wrist protectors to determine how much load reduction can be expected. The results of this analysis will help to develop performance guidelines. In summary, the results are:

- The bending stiffness is insufficient for protection with the short splint versions, but the protective effect increases with the longer versions.
- The dorsal splint and the sandwich splint reduce the wrist angle at impact slightly more than the palmar splint.
- The use of additional palm padding is least effective for children. This observation is true for long and medium splints and for dorsal and palmar variations.
- For all three age groups, the dorsal splint and the sandwich splint provide the biggest relative reduction in maximum force at the wrist joint.

- For the child model, the short versions of all the protectors showed significantly smaller effects than the medium and long versions.
- The long splint versions show more relative impact reduction than the medium versions. Hence, the long dorsal and sandwich splints provide the best protection, and the maximum protection is observed for the adult model.
- The splints that were modeled with padding show additional wrist-load reduction compared to splints modeled without padding.
- No significant differences were observed between 5 mm and 10 mm palm padding thicknesses.
- For the validated adult model in the reference condition, the maximum shear stress occurred at the palmar radius, just above the joint space. Using the protector produces maximal shear stress above the styloid process of the radius. A second maximum is observed in the shaft, which also shows increased shear stress.
- For the adolescent and child models, the long dorsal protector reduces the maximum shear stress at the hyaline cartilage plate. Again, and combined with increased shear stress, a second maximum was observed in the shaft.
- In the adolescent model, the palmar splint provided the highest stress reduction. In the child model, there was no significant difference between the dorsal and palmar splints.

With developed computer models, it is possible to acquire more knowledge about the functional requirements of wrist protectors, particularly for protecting children and adolescents.

Exercise

The following questions are intended to reflect the topic and may help to deepen the understanding for modelling and simulation.

1. To which category of models and which scientific discipline does the given example belong?
A: Category: Musculoskeletal models. Scientific discipline: Biomechanics

2. What is/are the reason(s) for implementing "wobbling masses" to a segment model of the human body?
A: If segments are modeled as stiff bodies, the calculated resulting joint forces will be unrealistically high for simulated movements with large accelerations. Large accelerations are typical for impacts, jumps or other highly dynamic movements.

OK. But why do wobbling masses reduce the joint forces?
A: Due to their visco-elastic coupling to the bone, they are accelerated or decelerated with a certain time delay relative to the bone. This means that, in the first instant of the impact, only the inertia of the skeleton and then a few milliseconds later the inertia of the rest of the segment mass are contributing

to the joint-reaction loads. For a drop jump, for instance, the difference in vertical knee-joint load is more than 250% between a stiff and a wobbling mass model.

Note

1 This investigation would not have been possible without the valuable contributions of a number of undergraduate and graduate students. The authors' special thanks go to (in alphabetic order): Dirk Baumeister, Franziska Damberger, Tobias Geyer, Nina Huber, Lennart Karstensen, Dominik Leitwein, Claudia Reinartz, Claudia Richter, Sebastian Schmidt, Lukas Stacheder, Florian G. Trittler und Ulrike Winkler.

References

Boehm, H. (2017, July). Modelling and *simulation of* the interaction between human and sport equipment to evaluate the functionality of sport technology. Habilitation thesis at the School of Sport and Health Sciences of Technical University of Munich.

Burkhart, T. & Andrews, D. (2010). The effectiveness of wrist guards for reducing wrist and elbow accelerations resulting from simulated forward falls. *J Appl Biomech*, 26(3), 281–289.

Chiu, J. & Robinovitch, S. (1998). Prediction of upper extremity impact forces during falls on the outstretched hand. *J Biomech*, 31, 1169–1176.

Deady, L. & Salonen, D. (2010). Skiing and snowboarding injuries: A review with a focus on mechanism of injury. *Radiol. Clin North Am*, 48, 1113–1124.

DeGoede, K. & Ashton-Miller, J. (2003). Biomechanical simulations of forward fall arrests: Effects of upper extremity arrest strategy, gender and aging-related declines in muscle strength. *J Biomech*, 36(3), 413–420.

DeGoede, K. & Ashton-Miller, J. (2002). Fall arrest strategy affects peak hand impact force in a forward fall. *J Biomech*, 35, 843–848.

European Committee for Standardization. (2007). Protective clothing: Wrist, palm, knee and elbow protectors for users of roller sports equipment. Requirements and test methods; EN 14120:2003+A1. EN 14120:2003+A1. Brussels: European Committee for Standardization.

Fiala, D., Psikuta, A., Jendritzky, G., Paulke, S., Nelson, D., van Marken Lichtenbelt, W. & Frijns, A. (2010). Physiological modeling for technical, clinical and research applications. *Frontiers in Bioscience S2*, June 1, 939–968.

Gebhardt, C. (2014). *Praxisbuch FEM mit ANSYS workbench: Einführung in die lineare und nichtlineare Mechanik*. Munich: Carl Hanser Verlag GmbH Co KG.

Greenwald, R., Janes, P., Swanson, S. & McDonald, T. (1998). Dynamic impact response of human cadaveric forearms using a wrist brace. *Am J Sport Med*, 26, 825–830.

Greenwald, R., Michel, F. & Simpson, F. (2013). Wrist biomechanics during snowboard falls. *Proceedings of the Institution of Mechanical Engineers, Part P: Journal of Sports Engineering and Technology*, 227(4), 244–254.

Gruber, K., Ruder, H., Denoth, J. & Schneider, K. (1998). A comparative study of impact dynamics: Wobbling mass model versus rigid body models. *J Biomech*, 31, 439–444.

Hagel, B., Goulet, C., Platt, R. & Pless, I. (2004). Injuries among skiers and snowboarders in Quebec. *Epidemiology*, 15, 279–286.

Hamer, A., Strachan, J., Black, M., Ibbotson, C., Stockley, I. & Elson, R. (1996). Biomechanical properties of cortical allograft bone using a new method of bone strength measurement. A comparison of fresh, fresh-frozen and irradiated bone. *J Bone Joint Surg [Br]*, 78-B, 363–368.

Idzikowski, J., Janes, P. & Abbott, P. (2000). Upper extremity snowboarding injuries: Ten-year results from the Colorado Snowboard Injury Survey. *Am J Sports Med*, 28(6), 825–831.

Kim, K., Alian, A., Morris, W. & Lee, Y-H. (2006). Shock attenuation of various protective devices for prevention of fall related injuries of the forearm/hand complex. *Am J Sport Med*, 34, 637–643.

Kim, S., Endres, N. & Johnson, R. (2012). Snowboarding injuries: Trends over time and comparisons with alpine skiing injuries. *Am J Sports Med*, 40(4), 770–776.

Kroncke, E., Niedfeldt, M. & Young, C. (2008). Use of protective equipment by adolescents in inline skating, skateboarding, and snowboarding. *Clin J Sport Med*, 18, 38–43.

Lehner, S. Entwicklung und Validierung biomechanischer Computermodelle und deren Einsatz in der Sportwissenschaft. Doctoral Dissertation at University Koblenz-Landau, Germany. http://kola.opus.hbz-nrw.de/volltexte/2008/266/ (accessed December 2014).

Lehner, S., Geyer, T., Michel, F., Schmitt, K-U. & Senner, V. (2014). Wrist injuries in snowboarding: Simulation of a worst case scenario of snowboard falls. *Engineering of Sport 10, Procedia Engineering*, 72, 255–260.

Leitwein, M. (2015). Numerical simulation of the human radius bone to examine typical snowboard injuries. Focuses: Optimization and validation of the model and analysis of results [in German]. Unpublished Bachelor's thesis at the Department of Mechanical Engineering of Technical University of Munich. Available on request from the authors.

Lewis, L., West, O., Standeven, J. & Jarvis, H. (1997). Do wrist guards protect against fractures? *Ann Emerg Med*, 29, 766–769.

Michel, F., Schmitt, K., Greenwald, R., Russell, K., Simpson, F., Schulz, D. & Langran, M. (2013). White paper. Functionality and efficacy of wrist protectors in snowboarding: Towards a harmonized international standard. *Sports Engineering*, 16(4), 197–210.

Moore, M., Popovic, N., Daniel, J., Boyea, S. & Polly, D. (1997). The effect of a wrist brace on injury patterns in experimentally produced distal radial fractures in a cadaveric model. *Am J Sport Med*, 25, 394–401.

Perrott, D., Richard, A., Smith, R., Leonard, B. & Kaban, L. (1992). The use of fresh frozen allogeneic bone for maxillary and mandibular reconstruction. *International Journal of Oral and Maxillofacial Surgery*, 21(5), 260–265.

Reilly, D. & Burstein, A. (1974). Review article. The mechanical properties of cortical bone. *J Bone Joint Surg Am*, 56(5), 1001–1022.

Russell, K., Hagel, B. & Francescutti, L. (2007). The effect of wrist guards on wrist and arm injuries among snowboarders: A systematic review. *Clin J Sport Med*, 17, 145–150.

Sasaki, K., Takagi, M., Ida, H., Yamakawa, M. & Ogino, T. (1999). Severity of upper limb injuries in snowboarding. *Arch Orthop Trauma Surg*, 119, 292–295.

Schmitt, K-U, Michel, F. & Staudigl, F. (2011). Analysing the impact behaviour of recent snowboarding wrist protectors. In K. Schmitt (Ed.), Proceedings of The IRCOBI Conference (pp. 51–61), September 14–16. Krakow, Poland.

Schmitt, K., Wider, D., Michel, F., Brügger, O., Gerber, H. & Denoth, J. (2012). Characterizing the mechanical parameters of forward and backward falls as experienced in snowboarding. *Sports Biomechanics*, 11, 57–72.

Staebler, M., Moore, D., Akelman, E., Weiss, A., Fadale, P. & Crisco, J. (1999). The effect of wrist guards on bone strain in the distal forearm. *Am J Sport Med*, 27, 500–506.

Varga, P., Baumbach, S., Pahr, D. & Zysset, P. (2009). Validation of an anatomy specific finite element model of Colles' fracture. *J Biomech*, 42(11), 1726–1731.

Wijdicks, C., Rosenbach, B., Flanagan, T., Bower, G., Newman, K., Clanton, T., ... Hackett, T. (2014). Injuries in elite and recreational snowboarders. *Br J Sports Med*, *48*(1), 11–17.

Woo, S., Orlando, C., Camp, J. & Akeson, W. (1986). Effects of postmortem storage by freezing on ligament tensile behaviour. *Journal of Biomechanics*, *19*(5), 399–404.

Wright, T. & Hayes W. (1976). Tensile testing of bone over a wide range of strain rates: Effects of strain rate, microstructure and density. *Medical and Biological Engineering*, *14*(6), 671–680.

Yamada, H. (1970). Strength of biological materials. In F. Gaynor Evans (Ed.). Baltimore, MD: Lippincott Williams & Wilkins Baltimore.

Zysset, P. (2009). Prediction of distal radius fracture load using HR–pQCT-based finite element analysis. *Dissertation*. www.ilsb.tuwien.ac.at/ilfb/pdfthesis/varga09.pdf (accessed January 2015).

12 Methods to gather key performance indicators for prosthetic feet

Felix Starker, Eric Nickel and Andrew Hansen

Introduction

Methods for sport-shoe testing can be classified in subjective, biomechanical and mechanical test methods (Odenwald, 2006). Today's frequently used mechanical test methods simplify load cases, so they can be performed by uniaxial test machines. Nevertheless, the results lack in quality when a more comprehensive understanding of the interaction of foot and shoe during the roll-over becomes important. Studies have shown (Starker, Blab, Dennerlein & Schneider, 2014) that test methods and procedures used to develop prosthetic feet can help as a pre-step between mechanical and complex biomechanical tests. Herein, the chapter focuses on prosthetic feet as a base model and ways to determine its performance indicators. A similar approach can be used to test, simulate and understand shoes and the shoe-foot interaction in a more comprehensive but repetitive way.

Prosthetic feet represent the final link between the prosthesis user and the ground. They support and help to propel the human body during the load transition from one leg to the other (Perry, 1992; Winter, 2009). The biological human foot is a complex part of the body consisting of 26 bones with several joint connections, ligaments, soft tissue and muscles allowing for both a robust impact resistant part and fine adjustability (Perry, 1992; Whittle, 1996; Winter, 2009). In contrast, most modern prosthetic foot designs attempt to mimic foot mechanics with a combination of materials with different stiffness characteristics providing spring and damping behavior with the aim of a robust, reliable design intended for a specific user weight, mobility and task. More complex tasks such as lifting the foot during swing phase to increase the ground clearance to prevent stumbling can be achieved with the addition of sensors and drive-system approaches. Nevertheless, the broad majority of today's prosthetic feet are mechanically passive.

The loss of a limb in developed countries is most often caused by peripheral vascular disease (Callaghan, Sockalingam, Treweek & Condie, 2002). Although media and manufacturers present healthy, active, young athletes, the majority of persons with amputation are of older age (Ziegler-Graham, MacKenzie, Ephraim, Travison & Brookmeyer, 2008; Treweek & Condie, 1998). With a wide variety of health and functional abilities of persons with amputations, there is a need to improve our understanding of prosthetic foot mechanics and to match the appropriate functional properties to each prosthesis user. The objective of this chapter is to provide recent methods that have been developed to measure key performance indicators (KPIs) of prosthetic feet, with specific focus on mimicking aspects of normal function with mechanical testing machines.

Project/problem

The human gait at normal walking speed can be described as a rolling motion of the foot around the ankle joint as well as additional joint axes within the foot (Kapandji, 2016). Prosthetic feet are designed to mimic this human behavior specifically during walking. Most modern prosthetic feet are passive mechanical systems made from carbon-fiber springs, cosmetic foams, metals and damping materials.

Even for experts, it can be difficult to distinguish between the different designs intended for the same prosthetic user category and estimate their performance.

State-of-the-art methods to investigate the performance of these devices is either through clinical studies with human subjects, through simplified mechanical tests (American Orthotic and Prosthetic Association, 2007; Smith & Gordon, 2017) or with the help of custom-made boots to investigate walking with prosthetic feet under load bearing by an able-bodied person (Zelik et al., 2011; Huang, Shorter, Adamczyk & Kuo, 2015).

Clinical studies and investigations in motion labs often have access to a limited number of similar prosthetic users for comparison. Additionally, these test users often have limited time for accommodation to a new foot design. The results are also a combination of prosthetic foot mechanics and the user's response to these mechanics. Standardized tests based on simple tensile testing devices capture solely mechanical parameters from quasi-static loading conditions that are not representative for a complete dynamic roll-over. Furthermore, the rigid attachment of test samples to machines can restrict the prosthetic foot sample in a non-biomechanical behavior, leading to unrealistic results. Latest mechanical testing approaches with subjects walking on a treadmill investigate both the mechanical characteristics with an external tuneable prosthetic foot (Zelik et al., 2011) and its metabolic consequences. Although closing the gap between foot performance and user preference, the simulated parameters allow only for a change in stiffness and additional push-off during late stance, but not a change in the mechanical design.

Other common activities, such as running, place different demands on the prosthetic leg. Higher impact loads can be observed in the springs, changing the mechanical properties and the performance. These changes can also be observed during outdoor activities such as hiking and climbing in rocky terrain. Due to difficult measuring conditions in the field, fewer studies have investigated the loading and unloading mechanics in these special environments (Oehler, 2016; Neumann, Brink, Yalamanchili & Lee, 2012; Frossard, Laurent, Beck, Dillon &

Evans, 2003). Transferring the study to the motion lab, obstacle courses need to be set up to simulate real environments.

Performance parameters for prosthetic feet

There are several properties of a prosthetic foot that have an impact on the performance of the prosthesis user. Within each specific foot design, there are many properties that could be considered; however, this chapter describes a "black-box" approach for examining prosthetic feet (Figure 12.1). This approach is chosen to allow measurement of many different types of systems and to allow useful comparisons between dramatically different designs. Input parameters can be measured both above and below the foot, e.g., by force, torque and acceleration sensors. Additionally, information of the shank position (angle, position) is required for a comprehensive analysis. Some of the most relevant "black-box" properties of prosthetic feet for walking that are a result from the input parameters are heel-loading impedance, energy stored and dissipated during gait, roll-over characteristics and the energy returned in late stance phase (Hubley-Kozey et al., 2017; Adamczyk, Collins & Kuo, 2006; Geil, 2001; Curtze et al., 2009).

The heel-loading impedance is an important characteristic for prosthetic feet as it helps to describe the initial shock absorption that the patient will feel during walking. If the impedance is too low, the foot will slap the ground and the ground reaction force will rapidly move under the rigid pylon connecting the foot with the knee. This low impedance provides a highly nonlinear loading that is likely uncomfortable and not ideal. If the heel impedance is too high, the user will suffer from limited shock absorption and the foot may not drop down to contact the floor until the midstance phase of walking. Depending

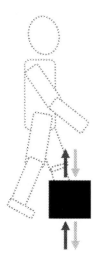

Figure 12.1 Input and output parameters describing a prosthetic foot.

on the user's weight, mobility and walking speed, the foot's impact angle and the impact energy differ (Klodd, Hansen, Fatone & Edwards, 2010). Elderly users tend to impact the floor with a reduced shank angle and can benefit from damping and reduction of the angular velocity due to reduced knee and hip stability. In contrast, highly mobile users land with a greater shank angle and, due to good balance, a fast transition to midfoot with minimal damping is more appropriate. The optimal spring-stiffness and damping therefore highly depends on the individual.

After this initial loading of the prosthetic foot, the person's center of mass then moves forward and over the foot. The foot deforms into a rocker shape allowing for a smooth roll-over transition from heel to toe (Wang & Hansen, 2010). This wheel-like rocker shape can either be a result from a stiff core located in the foot with a defined physical curvature or with springs deflecting under a load, creating a virtual curved shape. The energy stored and dissipated during this transition should be in correlation to the user's preference, activity and weight. For example, as users with low mobility tend to lose stability of the center of mass of the upper body during transition, they tend to take smaller steps. Furthermore, the resulting rocker shape of the foot, also referred as foot curvature, not only depends on the load the user puts into the foot but also on the user's leg length and body height.

Finally, when the opposing leg lands on the ground, the energy return the prosthetic foot provides in late stance determines how the load is transferred from one leg to the other (Russell Esposito, Aldridge Whitehead & Wilken, 2016; Klodd et al., 2010; Adamczyk et al., 2006). This high push-off force vector should occur at the appropriate angle to the ground and within a timing window that allows for a controllable transition.

Besides walking, the impact energy a prosthetic foot has to withstand and to absorb is of high relevance for highly active users. High-impact activities such as jumping or sudden acceleration/deceleration generate significant loads on the structure of the prosthesis that may lead to premature failure. The shock absorption the foot provides due to an appropriate impedance under specific loading conditions affects the load distribution of the soft tissue interface in the prosthetic socket.

Prosthetic feet can be studied in many different ways. Depending on the point of view and the sensors used within the study, the coordinate system may be either a fixed or moving reference frame.

In general gait or running, studies are performed with force plates fixed to the lab floor, and the prosthetic leg of the user is moving in the lab-based coordinate system. When measuring with sensors directly in the prosthesis, the coordinate system changes to a prosthesis or shank-based coordinate system (Figure 12.2). This consideration becomes relevant when forces and moments are compared between different methods and studies. Our further discussions of walking in this chapter are related to a shank-based coordinate system, which links effects directly to the user perception. The further discussion of impact resilience uses a lab coordinate system.

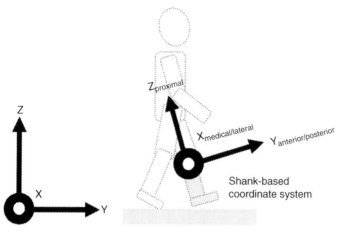

Figure 12.2 Different coordinate systems based on the location to measure data.

Concepts and methods of modelling

Modelling walking

Obviously, a method to investigate normal walking should simulate the motion profile of walking with its dynamic kinetics and kinematics as closely as possible. The kinematic chain of hip, knee and foot joints and ligaments is complex. Simplifying the knee joint as a rigid segment and the hip joint as a forgiving rotational element allows for a robust control of the system by a machine and allows for a free rotational motion around the hip joint. Moving the center of mass similar to the human upper body with a machine in space is difficult to simulate in a repetitive way (Starker, Blab, Dennerlein & Schneider, 2015). Therefore, the test apparatus that simulates walking is designed to rotate around a pivot below the foot to simulate the floor, to keep the foot like a pendulum straight and upright and to push with a linear actuator on the prosthetic leg to simulate the dynamic body weight (DIN, 2016).

KPIs for prosthetic feet during walking are parameters that are most relevant for the user and therefore parameters for the foot designer to focus on. To be able to compare between different types of feet, the test method should occur in a reproducible manner. Additionally, for some foot designs, loading at realistic values is important due to the damping characteristics of many materials, especially those used to cushion the heel pad but also hydraulic mechanisms used in some foot design. Therefore, for normal walking, the stance phase should be fully simulated in about 0.6 s (Perry, 1992).

To give elements in the foot time to fully unload, or investigate the performance of a toe lift, a swing phase in which the foot is lifted completely off the

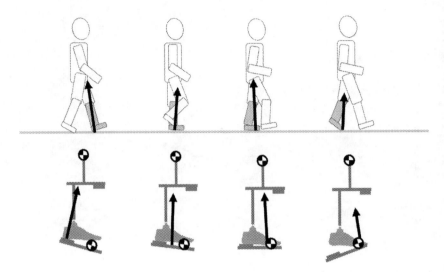

Figure 12.3 Translating the human walking motion in a simplified machine setup where the ground articulates to the foot.

ground is crucial. It is important to note that designs that act on gravitational effects (e.g., gyroscopes that detect the forward swing of the prosthetic leg) may not be adequately simulated with this approach.

To describe the relevant performance parameters, data from several sensors are required. The deformation of the foot can be captured either by optical systems or—in the case of a well-known reference frame—displacement sensors can be used. As prosthetic feet not only deform in the vertical direction but also provide rotational motion around an instantaneous center of rotation (Sawers & Hahn, 2011), it is important to record and process both displacements at the same time. Furthermore, forces and resulting moments can be captured through a force plate or a load cell in the limb assembly or within the floor.

The shank angle has to be measured with respect to the floor or—in the case of a machine setup like that shown in Figure 12.3—the plate angle that is recorded with respect to the machine frame. As timing effects are critical (e.g., for ankle moment vs. ankle angle), data synchronization between different sensors is crucial. Additionally, the sampling rate for all sensor signals should be sufficiently high (for walking, more than 100 Hz) to resolve fast-occurring events, for example during the transition from heel to toe during midstance.

As shoes have a large effect on the performance and mechanical parameters of prosthetic feet, the samples are aligned with a heel block at the foot plate to simulate the heel height of a shoe, but are then tested without shoes. The effects of different shoes can also be simulated on the feet by then removing the heel block (Major, Scham & Orendurff, 2018; Starker 2016).

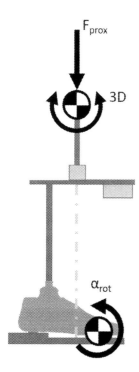

Figure 12.4 Setup of the prosthetic foot in a machine according to DIN (2016).

Methodological experience

Basics of the mechanical test method for walking simulation

A multiaxial test machine according to a standardized method (DIN, 2016) described allows for a deeper insight into the performance of a foot during walking for a defined dynamic load profile. The foot is aligned according to geometrical aspects to allow for a static vertical force vector running through half of the foot length, as shown in Figure 12.3. A ball joint with very low reset spring force above the overall limb setup ensures that the test sample is able to rotate freely under the given test loads. A six-component load cell close to the height of the knee joint captures forces and moments in all directions. Furthermore, a camera is mounted to the leg, looking at the side of the foot sample, tracking the actual deformation based on passive markers attached to both the prosthetic pylon and the cosmetic cover in a fixed coordinate system allowing for easy data extraction.

Compressive force is applied above the ball joint using a linear actuator. The floor is represented by a tilting table under the foot. This approach simulates walking, but with the foot in a nearly stationary coordinate system. To minimize effects of unrealistic torque and shear forces due to an overdetermined system, the leg is capable of pivoting slightly around the ball joint attached with low reset spring force.

Alignment of the foot sample

One of the key issues in comparing today's prosthetic studies is the high variability of the setup of prosthetic foot samples, even in mechanical tests, due to the prosthetic connection allowing for rotation along a ball surface on each connector included in the prosthetic leg (Fridman, Ona & Isakov, 2003; Kobayashi, Orendurff, Arabian, Rosenbaum-Chou & Boone, 2014). To eliminate the high variability of alignment, a dynamic approach is used to set the alignment of the foot samples. A complete roll-over with half of the body weight is performed, which is comparable to the load when standing on two legs. The foot is manually shifted after each trial so that the transition point of the moment in the sagittal plane passes at 0° of the tilt angle of the floor with respect to the horizontal.

Performance parameters of a prosthetic foot

Walking motion and implications to key performance indicators

The motion of a prosthetic foot can be described with both a deformation θ (d_{foot}) and rotation ($\theta_{heel/ankle}$). Because forefoot and heel (θ_{heel}) might act independently of each other, the actual ankle angle should be measured from the rigid prosthetic connector on top of the foot to a point below and from this point to a point where the forefoot spring ends close to the toe section of the foot. (Figure 12.4) Furthermore, the rocker-curve ($s_{x,y}$) can be calculated by both the deformation of the foot and the center of pressure which is the resulting point of the force vector through the sole of the foot.

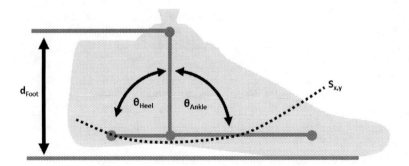

Figure 12.5 Displacement (dFoot) and rotation (θ Heel/Ankle) occurring at the prosthetic foot.

Results

Simulating walking

The KPIs described are linked directly with the mechanical parameters of the prosthetic foot as a combination of results from a series of test scenarios.

In the following section, the KPIs are described with two different prosthetic feet intended for different user groups based on a roll-over test simulating a person with 70 kg of body weight according to (DIN, 2016). Foot no. 1 (left side) is intended for less mobile users that ambulate primarily in a home environment and sometimes out of their house with reduced walking speed. Foot no. 2 (right side) is intended for moderate-activity people with normal walking speed as well as moderate outdoor activities.

Time and timing

Time and timing is a coupled factor of parameters such as walking speed, shank angle vs. ground angle and the center of mass relative to the prosthetic foot, resulting in a bending moment at the ankle joint.

Forces

For the vertical force, two characteristic maximum points can be found during normal walking speed (5 to 6 km/h). The first peak is a result from the collision of the heel and the ground with about 120% of the body weight. The second peak in the force curve results from the transition of the body weight to the other leg. During midstance, the center of mass of the body has returned upward from a lowered position resulting in reduced load on the leg.

The anterior/posterior forces are an effect from the transition of the force vector through the foot.

$$F_{x,y,z}[N]$$

Moments

The resulting moment measured either at the load cell above or at a force plate underneath the foot results from the mechanical structure forming heel and toe and its resilience during the deformation. Not only shape and maximum values but also transition of the moment in the sagittal plane (Figure 12.7) represents the transition from the heel to the toe structure. As a defined joint center is hard to select in a prosthetic foot, the moments are calculated down to the floor level.

$$M_{x,y,z}[Nm]$$

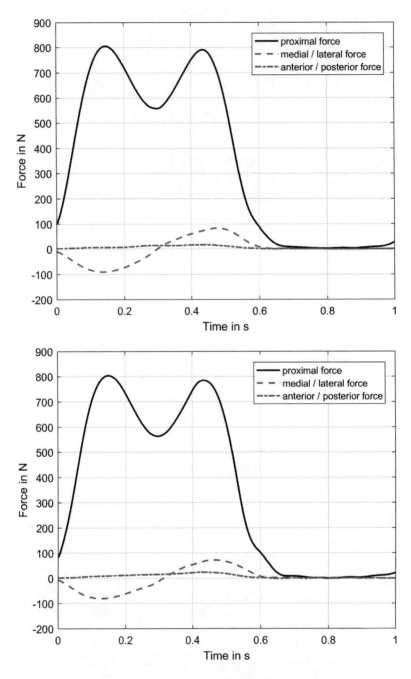

Figure 12.6 Force curves of two test samples. The proximal pointing curve (black) is controlled by the test machine and given by the test standard DIN (2016). Nevertheless, characteristic differences can be observed in midstance as well as the anterior/posterior forces (gray).

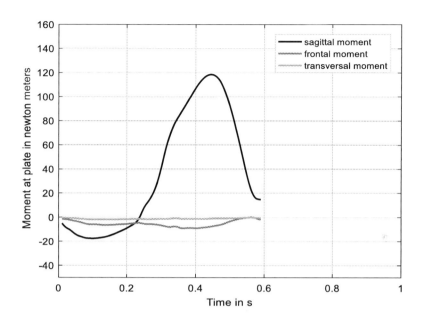

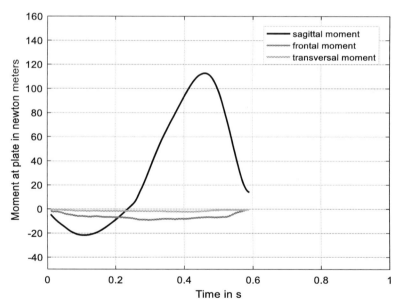

Figure 12.7 The highest moment of the prosthetic foot occurs in the sagittal plane. First, a counterclockwise bending moment can be observed. During the transition of the foot from heel to midfoot, a very low moment is present. While rolling forward on the toes, the center of pressure moves further to the toes, resulting in a high clockwise bending moment with a typical magnitude of 120 to 150 Nm for a 70 kg person.

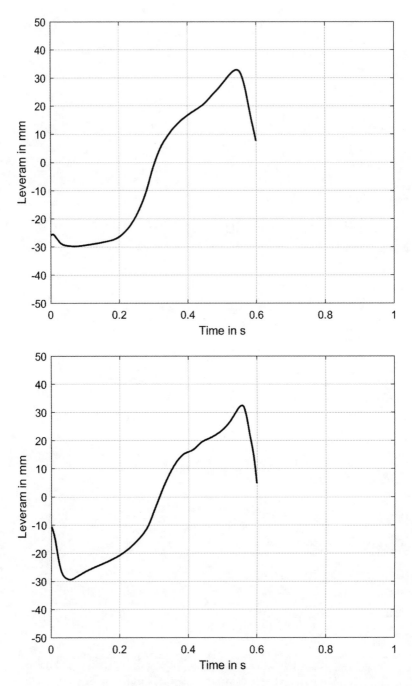

Figure 12.8 The effective lever arm not only differs in maximum values between different feet but also describes the time and duration of transition from heel to midfoot, full ground contact and, finally, heel-off. Sharp edges in this curve describe a rough progression of the force vector.

Supporting lever arm

Dividing the sagittal moment by the vertical force results in the so-called "supporting lever arm" (Figure 12.8) A short lever arm can be interpreted as a short keel of the foot that might be easy to overcome for a less active patient with shorter, less dynamic steps, but that might create a "drop-off" effect in late stance phase for persons that walk above certain speeds. The lever arm can be calculated directly from the moment and the vertical force:

$$l_{x,foot} = \frac{M_{x,floor}}{F_z} [mm]$$

Roll-over shape

The roll-over shape describes the resulting deformation curve of each segment of the prosthetic foot during the complete heel-to-toe roll-over (Hansen & Childress, 2002; Miff, Hansen, Childress, Gard & Meier, 2008; Curtze, Otten, Hof & Postema, 2011). This deformation profile shows the "effective rocker shape" of the foot during walking. The overlay of a timed point series of the center of pressure helps to understand the timing of the transition during the roll-over (Figure 12.9). For a smooth progression of the force vector, the points are evenly distributed. In contrast to some prosthetic feet for less active users, these points are located at two positions—one at the heel and one in the forefoot section—meaning the foot initially stays at the heel and suddenly rolls forward to the toes and remains there until it is lifted off the ground.

Depending on the test setup, the roll-over shape is calculated differently (Hansen & Meier, 2010; ISO, 2016) and represents a trajectory in space. Nevertheless, the most biomechanically understood and relevant is its projection into the sagittal plane.

$$s_{x,y} = (l_{x,foot}, d_{Foot})[mm]$$

Impedance

The impedance of the foot can be examined with the same test approach and machine but with static floor positions with a loading/unloading pattern to determine the impedance (stiffness) of the foot at specific angles. Measuring the vertical force vs. the vertical displacement results in the quasi-stiffness of the affected foot segment under different loading angles (Figure 12.10). The slope of this curve represents the impedance/stiffness; the resulting hysteresis curve with its surface area represents the energy loss during loading and unloading.

$$I_{total} = \frac{\Delta F_{min\ to\ max}}{\Delta d_{0\ to\ max}}$$

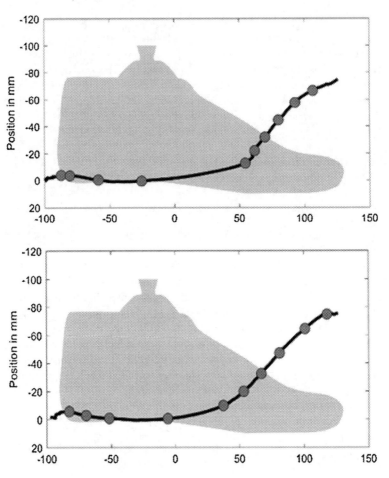

Figure 12.9 The shape of the effective rocker can be simplified as a circle radius during the support stance phase. In this figure, the rocker shape is shown throughout the full stance phase of the foot and therefore may differ from known shapes reported in the literature (Hansen, Childress & Knox, 2000; Curtze et al., 2009). The roll-over shape of prosthetic feet differs between foot designs, the total load and speed. A long progressing curvature allows for a long rolling motion over the "toes" of the foot. This can be beneficial for faster waking speed and a good load transition.

The energy dissipated can be calculated with the following formula:

$$E_{\textit{eff}} = \int_{Fmin\_1}^{Fmax} F_z \, ds - \int_{Fmax}^{Fmin\_2} F_z \, ds$$

Typical values for prosthetic foot spring and damper systems range from 50% to 96% effective energy return (Geil, 2001; Postema, Hermens, de Vries, Koopman & Eisma, 1997; Takahashi & Stanhope, 2013).

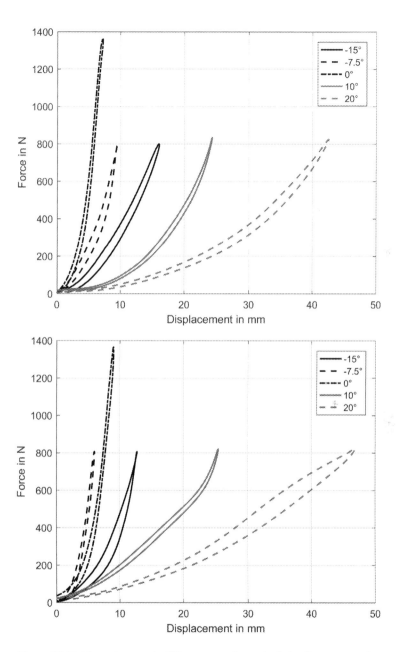

Figure 12.10 Measuring quasi-stiffness curves from single loading tests at given test loads under defined static floor angles reveals the energy storage and return without clarifying whether the motion occurred in a deformation or a rotation.

Stiffness values for a static loading test at 12° plantar flexion angle reported for prosthetic feet range between 0.0277 N/m to 0.076 N/m (Geil, 2001). For the sample data presented here, stiffness values of 0.029 N/m to 0.039 N/m were found for 10° of plantar flexion angle.

Ankle angle vs. ankle moment

In traditional linear displacement analyses, it is difficult to link energy return results directly to user performance. The relationship between energy return and user performance is clearer when considered in a rotational sense, using the relationship between ankle torque and ankle angle (Rouse, Gregg, Hargrove & Sensinger, 2013). The rotation angle of the foot ($\theta_{Heel/Ankle}$) is captured by the camera. Calculating moments from the center of pressure during the loading, and combining these moments with the angle data from the cameras, results in the deformation solely due to the rotation. The remaining deformation occurs necessarily due to actual linear displacement (Figure 12.11).

$$d_{foot,total} = d_{foot,rotation} + d_{foot,linear} = Tan(\theta) * s_{COP,y} + d_{foot,linear}$$

The rotational stiffness can be calculated by the slope in the quadrant 3 and quadrant 1 of the ankle angle vs. ankle moment curve in Figure 12.11.

In conclusion, there are several approaches that can be used to determine KPIs of prosthetic feet. Based on the presented test setup, data is reproducible with the limitation that input parameters (vertical force and shank angle profile) are predetermined.

For walking, static loading and unloading of the heel can be used to examine heel impedance properties. The roll-over shape can be used to examine the overall "rolling" feel of the foot for the user. Lastly, the energy returned in late stance phase seems important to energetics of walking.

Although evidence in clinical studies of specific prosthetic populations seems to be linked to these KPIs, more work is required to describe and understand the individual demands better and finally to select and provide the optimal prosthetic foot for each individual user.

Example

Exercise

Within the following exercise based on a standardized foot of 260 mm length, several foot parameters shall be calculated (Figure 12.12).

1) Calculate based on the specific points (-15° to +20°) in Figure 12.11 the deformation of the foot in a vertical direction excluding the deformation of the foot occurring due to a plantar flexion of the foot (rotational motion).

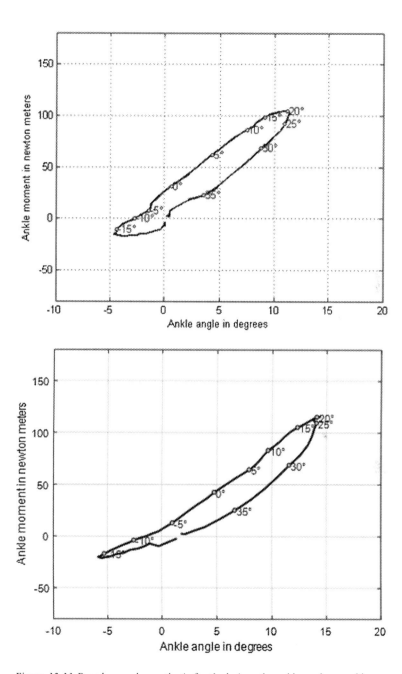

Figure 12.11 Based on each prosthetic foot's design, the ankle angle vs. ankle moment not only represents the range of motion it undergoes during a full roll-over under a standardized input angle slope (-20° to +40°), but also how much energy during this rotation is dissipated. The slope of the curve is directly linked to the angular stiffness during the transition. During heel strike, the negative bending moment results in a plantar flexion of the foot which helps for a deceleration of the body weight.

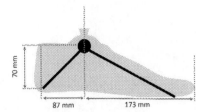

Figure 12.12 Dimensions of a prosthetic foot used for the example.

Table 12.1 Specific parameters extracted from test data presented earlier.

	Foot 1		Foot 2	
Floor angle [°]	−15	+20	−15	+20
d_{Foot} [mm]	12.4	40.8	12.8	39.5
$\Theta_{Heel}/\Theta_{Toe}$ [°]	−4.3	10.5	−5.9	14.1
$S_{COP,y}$ [mm]	−87.2	70	−82.4	81.2
M_{Ankle} [Nm]	−17.5	118.3	−21.7	112.7

According to the formulas

$$d_{foot,rotation} = \tan\theta * COPy$$

$$d_{foot,linear} = d_{foot} - d_{foot,rotation}$$

the results for both feet calculate:

	Foot 1		Foot 2	
$d_{foot,rotation}$ [mm]	6.6	13	8.5	20.4
$d_{foot,linear}$ [mm]	5.8	27.8	4.3	19.1

2) Calculate the rotational stiffness of both prosthetic feet and conclude which design might be more suitable for less stable prosthetic users.

Based on the results from the table, we can calculate the rotational impedance/stiffness with

$$I_{rotation} = \frac{M_{min/max}}{\theta_{heel/toe}}$$

	Foot 1		Foot 2	
$I_{rotation}$ [Nm/°]	4.1	11.3	3.7	8.0

3) Calculate the effective rocker radius based on the assumption of an ideal circle.

Based on three points (X, Y) of the COP, e.g., at -10°, 0°, 10° in Figure 12.8, we can fit a circle and estimate the radius.

	Foot 1			Foot 2		
	-10°	0°	+20°	-10°	0°	+20°
S_{COPx} [mm]	-80.1	-25.4	62	-69	-6.1	53.3
S_{COPy} [mm]	3.4	-0.5	21.8	2.7	0.4	19.8

Using the circle equation

$$(x - x_0)^2 + (y - y_0)^2 = r^2$$

the resulting radius calculates as the following:

	Foot 1	Foot 2
r [mm]	225	177

Modelling of downward jumping

The human musculoskeletal system is exceptionally resilient to stresses when compared to synthetic materials (Klute & Berge, 2004). Combined with the human capacity to repair minor damage, it is a significant challenge for engineers to design components for lower-limb prostheses within the mass, cosmetic envelope, durability and functional performance constraints. Current test standards for prosthesis durability address static overload structural strength (using uniaxial compressive forces along a specific load line intended to simulate the loading conditions in early or late stance) and cyclic fatigue strength (repetitive heel and forefoot loading or heel-to-toe rolling loads) over millions of cycles.

For most lower-limb prosthesis users, the current test standards (ISO/TC 168 Prosthetics and Orthotics, 2016a; ISO/TC 168 Prosthetics and Orthotics, 2016b) address the loads induced by gait and standard daily activities. Active prosthetic users have different functional expectations, whether it be the needs of a sport (e.g., running) or an active profession (e.g., farming). For athletes, highly specialized prostheses can be provided for a specific sport. For prosthesis users who have an active profession, their normal prosthesis needs to provide for the functional demands of their daily life, yet be durable enough for their workplace activities. To promote safety in the workplace and to avoid product failure under anticipated use cases, prostheses meant for high-activity users should be tested under realistic impact loading conditions.

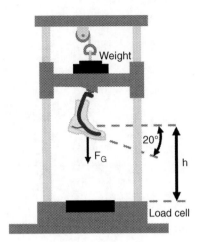

Figure 12.13 Schematic of a drop-test setup.

Methodological experience[1]

For these higher-impact situations, such as jumping off an obstacle, a different test method from simulating walking is required. This special task is not expected to occur frequently, but failure of the prosthesis under these anticipated use cases could result in injury or great expense for the user. Compared to walking impact energy, energy dissipation and structural integrity after overload becomes of higher relevance.

For the machine setup, the prosthetic foot is aligned plantar flexed and a deadweight is added above. Guides allow for linear motion. The drop height is specified and executed at different levels. Forces are measured with a load cell within the floor below the sample (as shown in Figure 12.13) or connected between the sample and the mass.

For impact testing, the shoes to be worn can have a great effect on the system performance and the combined investigation of the impact effects of shoe and prosthetic foot is of high relevance.

Impact testing and implications to key performance indicators

To simulate the impact during jumping activities, a simplified mass, spring and damper model arrives at a quick approximation (Figure 12.14). The mass represents the overall body mass. Damping factors can be a result from the prosthetic foot (e.g., shock absorber) or a bending of joints of the body such as the hip and the knee joint. The example for the modelling approach described later also points out the different demands of different user groups. A below-knee amputee is most likely able to use hip and knee flexion to decelerate the impact. In contrast, an above-knee amputee has to keep the prosthetic knee at a straight angle to prevent knee collapse. With a straight knee, the impact forces are higher

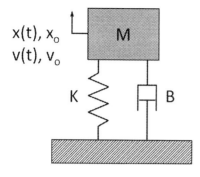

Figure 12.14 Linear spring-mass-damper diagram.

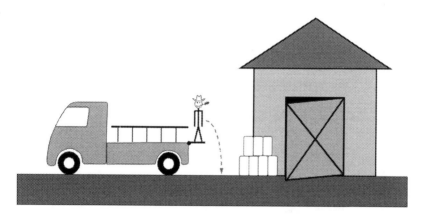

Figure 12.15 Schematic of an example for downward jumping.

but can be compensated for by an additional element in the foot such as a shock absorber, although these components usually do not have large travel.

The characteristic equation of motion of a linear spring-mass-damper system with downward directed gravitational force can be estimated as:

$$\ddot{x} + 2\xi\omega_o\dot{x} + \omega_o^2 x = -g$$

where $\xi = \dfrac{B}{2\sqrt{KM}}$ and $\omega_o = \sqrt{\dfrac{K}{M}}$

Example

This real-life example explains the modelling of the impact resilience of a prosthetic foot (Figure 12.15). It is based on an infrequent but occurring event not covered by any standard test (ISO/TC 168 Prosthetics and Orthotics, 2016a;

ISO/TC 168 Prosthetics and Orthotics, 2016b; AOPA, 2007). A farmer with 100 kg of body weight has finished unloading bags of potatoes from his truck. The farmer needs to get down, so he can put the bags into storage. He jumps from the back of the truck to get down. The back of the truck is 0.70 m above the ground and the farmer drops down as directly as possible (without significant forward movement).

Task 1: Calculate the peak impact force on the farmer's legs.

Velocity at impact

Assuming a perfect conversion of potential energy to kinetic energy, at impact, the farmer will have a velocity of:

$$E_{kin} = E_{pot}$$

$$\frac{mv^2}{2} = mgh$$

$$v = \sqrt{2gh}$$

$$v = \sqrt{2*9.81 \left[\frac{m}{s^2}\right]*0.7\,[m]} = 3.71\,[\frac{m}{s}]$$

(Note that due to the square root, this value for the velocity does not indicate the direction, only the magnitude and is therefore the speed. The sign must be determined by defining the direction of positive velocity when setting up the impact mechanics model.)

Impact mechanics

When the farmer lands on the ground, he uses his legs as a spring and damper to absorb the impact and restore himself to a neutral position.

In this case, the farmer contacts the ground at x_o = 0.0 m, but due to the free fall from the truck, he has an initial velocity of v_o = -3.71 m/s (the magnitude, calculated previously, is in the downward direction but, in the model above, the velocity is defined as positive in the upward direction, therefore the sign on the initial velocity is negative).

Thus, for the model above, the second-order spring-mass-damper equation of motion takes the form:

$$\ddot{x} = -\frac{B}{m}\dot{x} - \frac{K}{m}x - g \text{ where } x_o(0) = 0\,m \text{ and } \dot{x}_o(0) = -3.96\,m/s$$

Assuming the farmer's legs provide a net spring constant K = 30,000 N/m and damping of B = 2,000 Ns/m, the initial acceleration at contact can be calculated:

$$\ddot{x}_o = -\frac{2000\left[\mathrm{Ns/m}\right]}{100[kg]}\dot{x}_o - \frac{30{,}000\left[\mathrm{N/m}\right]}{100[kg]}x_o - 9.80665\left[\mathrm{m/s^2}\right]$$

$$\ddot{x}_o = -\frac{2000\left[\mathrm{Ns/m}\right]}{100[kg]}(-3.71\left[\mathrm{m/s}\right]) - \frac{30{,}000\left[\mathrm{N/m}\right]}{100[kg]}(0[m]) - 9.80665\left[\mathrm{m/s^2}\right]$$

$$\ddot{x}_o = 64.3\left[\mathrm{m/s^2}\right]$$

Using these initial values, the time course of the impact load response can be calculated. The displacement and ground reaction force for the farmer's landing are shown in Figure 12.16.

The vertical ground reaction force (VGRF) shown in Figure 12.5 peaks at approximately 8,400 [N]. This force is distributed to both legs, so each leg experiences approximately 4,200 [N] peak force during the impact. Note that this force equals the 100 [kg] farmer carrying additional load of 750 [kg].

For a farmer who uses a lower-limb prosthesis, the problem gets a little more challenging. We do not know how much the farmer can use his hip flexion and knee flexion to absorb the impact, and it may be dramatically different from a person who does not have an amputation. Data on amputees jumping down from an elevation is difficult to obtain for safety reasons (the risk of an injury to the skin, a fall or damage to the prosthesis), therefore, we need to make some assumptions to begin to create a test system for prostheses to evaluate performance under impact loading. The worst possible case for the prosthesis (and probably for the farmer as well) would be to have very little or no knee and hip flexion (very stiff leg, essentially directly coupling the body mass to the prosthesis). Also, it is unlikely that the farmer would intend to land with more weight on the prosthesis than his other leg, so, since we are already being conservative by not including knee and hip flexion, we can assume equal weight distribution, thus only half of the farmer's weight is landing on the leg with the prosthesis.

The current test system being used to evaluate prostheses under impact loading can record the vertical impact force, the velocity at impact and whether the prosthesis suffered a failure. The simplest outcome of impact testing is at what drop height the sample failed. However, there are many parameters that could affect the results. Depending on the foot length, the lever arm of the prosthetic foot changes and can alter the internal mechanics of the structural loads. The body weight for which the prosthesis was intended affects the stiffness of the spring elements and should affect the load applied to the sample during the drop. The

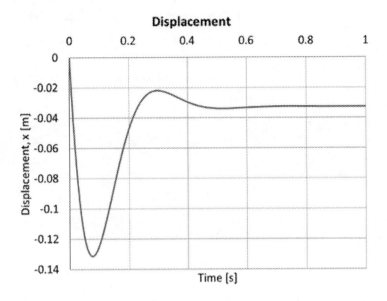
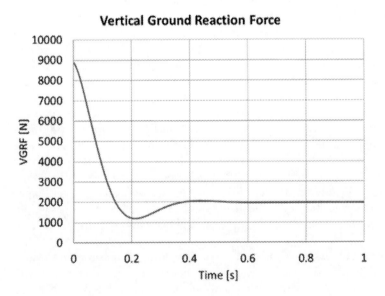

Figure 12.16 Impact displacement and vertical ground reaction force for the prosthetic foot model.

intended use cases would affect what drop height would be considered "acceptable" or "passing." These factors must be accounted for when considering the approach to testing.

The velocity measurement is primarily a means to determine whether the system is truly approximating free fall at impact and, if not, it can be used to determine an effective free-fall drop height (the height at which free fall would produce the measured velocity). Assuming the impact test is a perfect free fall, and assuming the farmer lands with locked legs (very uncomfortable for the farmer and likely to damage his knees, but the worst-case scenario for the prosthesis), we can obtain force profiles of the impact. An example of this force profile for a prosthetic foot loaded with a mass of 45.9 [kg] and dropped from a height of 70 cm is shown in Figure 12.17.

Looking closer at the initial impact, (Figure 12.18) we can see that the impact begins at t = 0.857 and ends at t = 0.948 with peak force reached at approximately t = 0.896 (for a rise time of 0.039 [s]). A closer look at the later impacts (Figure 12.19) shows a non-sinus-shape force profile. The simple linear spring fails in prediction, but it is valid for a first assumption.

From the data, we can calculate the natural frequency of oscillation (ω_o) and the damping ratio (ξ) which, in turn, will let us calculate the stiffness and damping coefficients (linear approximations of the constants K and B. See Figure 12.20).

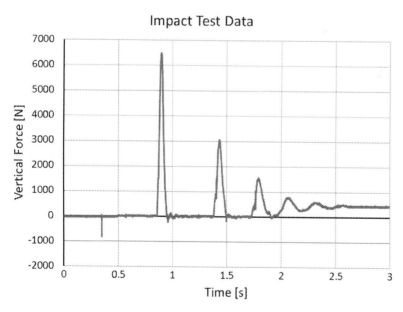

Figure 12.17 Impact forces vs. time of a drop test of a prosthetic foot.

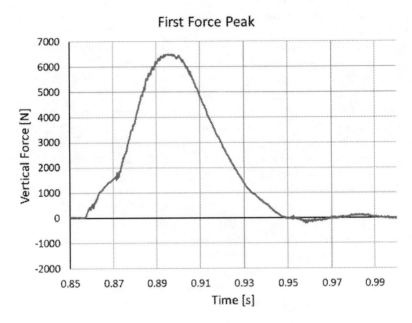

Figure 12.18 First impact force vs. time of a prosthetic foot.

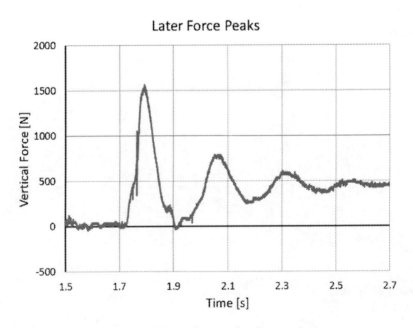

Figure 12.19 Spring mass oscillations (later peaks of impact data).

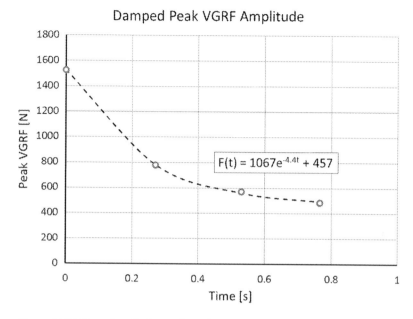

Figure 12.20 Plot of oscillation peaks with exponential fit curve.

The average period between peaks during oscillation was 0.255 [s]. Only the peaks where the foot remained on the ground continuously between peaks may be used, otherwise the ballistic period where the foot was off the ground must be accounted for, complicating the calculations. Using the average value for the oscillation period, the natural frequency is $\omega_o = 2\pi / 0.255s = 24.6\,[s^{-1}]$.

The exponent on the decay of the VGRF peaks during oscillation is -4.4 and is equal to $-\xi\omega_o$. Having already determined ω_o to be 24.6, the damping ratio is $\xi = 0.179$. We can then use $\xi = \dfrac{B}{2\sqrt{KM}}$ and $\omega_o = \sqrt{\dfrac{K}{M}}$ to solve for K and B. In this example, these values are K = 27,800 [N/m] and B = 404 [Ns/m]. We also can directly measure the first peak, which is the maximum applied force. For this test, the peak force was 6,480 [N].

Note

1 Support for development of the impact loading machine was provided by the BADER Consortium via the Congressionally Designated Medical Research Program (Award # W81XWH-11-2-0222). Opinions, interpretations, conclusions and recommendations are those of the authors and are not necessarily endorsed by the Department of Defense, Department of Veterans Affairs, Department of the Army or Department of the Navy.

References

Adamczyk, P., Collins, S. & Kuo, A. (2006). The advantages of a rolling foot in human walking. *The Journal of Experimental Biology*, *209*(20), 3953–3963.

American Orthotic and Prosthetic Association (AOPA). (2007). AOPA Prosthetic foot project. *Patient Care*, 44.

Callaghan, B., Sockalingam, S., Treweek, S. & Condie, M. (2002). A post-discharge functional outcome measure for lower limb amputees: Test-retest reliability with trans-tibial amputees. *Prosthetics and Orthotics International*, *26*(2), 113–119.

Curtze, C., Hof, A., van Keeken, H., Halbertsma, J., Postema, K. & Otten, B. (2009). Comparative roll-over analysis of prosthetic feet. *Journal of Biomechanics*, *42*(11), 1746–1753.

Curtze, C., Otten, B., Hof, A. & Postema, K. (2011). Determining asymmetry of roll-over shapes in prosthetic walking. *The Journal of Rehabilitation Research and Development*, *48*(10), 1249.

DIN (2016). ISO/TS 16955:2016 Prosthetics – Quantification of physical parameters of ankle foot devices and foot units. Geneva: ISO.

Fridman, A., Ona, I. & Isakov, E. (2003). The influence of prosthetic foot alignment on trans-tibial amputee gait. *Prosthetics and Orthotics International*, *27*, 17–22.

Frossard, L., Laurent, A., Beck, J., Dillon, M. & Evans, J. (2003). Development and preliminary testing of a device for the direct measurement of forces and moments in the prosthetic limb of transfemoral amputees during activities of daily living. *JPO Journal of Prosthetics and Orthotics*, *15*(4), 135–142.

Geil, M. (2001). Energy loss and stiffness properties of dynamic elastic response prosthetic feet. *JPO Journal of Prosthetics and Orthotics*, *13*(3), 70–73.

Hansen, A. & Childress, D. (2002). Roll-over characteristics of human walking with applications for artificial limbs. 3050531 (June), 332.

Hansen, A., Childress, D. & Knox, E. (2000). Prosthetic foot roll-over shapes with implications for alignment of trans-tibial prostheses. *Prosthetics and Orthotics International*, *24*(3), 205–215.

Hansen, A. & Meier, M. (2010). Roll-over shapes of the ankle-foot and knee-ankle-foot systems of able-bodied children. *Clinical Biomechanics*, *25*(3), 248–255.

Huang, T., Shorter, K., Adamczyk, P. & Kuo, A. (2015). Mechanical and energetic consequences of reduced ankle plantar-flexion in human walking. *Journal of Experimental Biology*, *218*(22), 3541–3550.

Hubley-Kozey, C. & Wilson, J. (2017). Effects of knee osteoarthritis and joint replacement surgery on gait. In B. Müller & S. Wolf (Eds.), *Handbook of human motion*. Cham: Springer International Publishing.

ISO (2016). *ISO TS 16955: Prosthetics — Quantification of physical parameters of ankle foot devices and foot units*. Geneva: ISO.

ISO/TC 168 Prosthetics and Orthotics (2016a). ISO 10328:2016 Prosthetics: Structural testing of lower-limb prostheses – Requirements and test methods, 2016.

ISO/TC 168 Prosthetics and Orthotics (2016b). ISO 22675:2016 Prosthetics: Testing of ankle-foot devices and foot units – Requirements and test methods, 2016.

Kapandji, A. (2016). *Funktionelle anatomie der gelenke* (6th edn.). Stuttgart: Thieme.

Klodd, E., Hansen, A., Fatone, S. & Edwards, M. (2010). Effects of prosthetic foot forefoot flexibility on gait of unilateral transtibial prosthesis users. *Journal of Rehabilitation Research and Development*, *47*(9), 899–910.

Klute, G. & Berge, J. (2004). Modelling the effect of prosthetic feet and shoes on the heel-ground contact force in amputee gait. *Proceedings of the Institution of Mechanical Engineers, Part H: Journal of Engineering in Medicine*, *218*(3), 173–182.

Kobayashi, T., Orendurff, M., Arabian, A., Rosenbaum-Chou, T. & Boone, D. (2014). Effect of prosthetic alignment changes on socket reaction moment impulse during walking in transtibial amputees. *Journal of Biomechanics*, 47(6), 1315–1323.

Major, M., Scham, J. & Orendurff, M. (2018). The effects of common footwear on stance-phase mechanical properties of the prosthetic foot-shoe system. *Prosthetics and Orthotics International*, 42(2), 198–207.

Miff, S., Hansen, A., Childress, D., Gard, S. & Meier, M. (2008). Roll-over shapes of the able-bodied knee-ankle-foot system during gait initiation, steady-state walking, and gait termination. *Gait and Posture*, 27(2), 316–322.

Neumann, E., Brink , J., Yalamanchili , K. & Lee, J. (2012). Use of a load cell and force-moment analysis to examine transtibial prosthesis foot rollover kinetics for anterior-posterior alignment perturbations. *JPO Journal of Prosthetics and Orthotics*, 24(4), 160–174.

Odenwald, S. (2006). Test methods in development of sport equipment. In E. Moritz & S. Haake (Eds.), *The Engineering of Sport 6* (pp. 301–306). New York: Springer.

Oehler, S. (2016). *Mobilitätsuntersuchungen und belastungsmessungen an oberschenkelamputierten*. Berlin: De Gruyter.

Perry, J. (1992). *Gait analysis: Normal and pathological function*. Thorofare, NJ: Slack.

Postema, K., Hermens, H., de Vries, J., Koopman, H. & Eisma, W. (1997). Energy storage and release of prosthetic feet Part 1: biomechanical analysis related to user benefits. *Prosthetics and Orthotics International*, 21, 17–27.

Rouse, E., Gregg, R., Hargrove, L. & Sensinger, J. (2013). The difference between stiffness and quasi-stiffness in the context of biomechanical modeling. *IEEE Transactions on Biomedical Engineering*, 60(2), 562–568.

Russell Esposito, E., Aldridge Whitehead, J. & Wilken, J. (2016). Step-to-step transition work during level and inclined walking using passive and powered ankle-foot prostheses. *Prosthetics and Orthotics International*, 40(3), 311–319.

Sawers, A. & Hahn, M. (2011). Trajectory of the center of rotation in non-articulated energy storage and return prosthetic feet. *Journal of Biomechanics*, 44(9), 1673–1677.

Smith, K. & Gordon, A. (2017). Mechanical characterization of prosthetic feet and shell covers using a force loading apparatus. *Experimental Mechanics*, 57(6), 953–966.

Starker, F. (2016). Untersuchung des Einfluss von Schuhen auf das mechanische Verhalten von Prothesenfüßen mit einer Prothesenfußprüfmaschine. OTWorld, May 3–6, 2016, Leipzig, Germany.

Starker, F., Blab, F., Dennerlein, F. & Schneider, U. (2014). A method for sports shoe machinery endurance testing: Modification of ISO 22675 prosthetic foot test machine for heel-to-toe running movement. *Procedia Engineering*, 72, 405–410.

Starker, D., Blab, F., Dennerlein, F. & Schneider, U. (2015). Simulation des Treppen- und rampengangs mit einer Prothesenfusstestmaschine, May 6–8, 2015, Bonn, Germany.

Takahashi, K. & Stanhope, S. (2013). Mechanical energy profiles of the combined ankle-foot system in normal gait: Insights for prosthetic designs. *Gait and Posture*, 38(4), 818–823.

Treweek, S. & Condie, M. (1998). Three measures of functional outcome for lower limb amputees: A retrospective review. *Prosthetics and Orthotics International*, 22(3), 178–185.

Wang, C. & Hansen, A. (2010). Response of able-bodied persons to changes in shoe rocker radius during walking: Changes in ankle kinematics to maintain a consistent roll-over shape. *Journal of Biomechanics*, 43(12), 2288–2293.

Whittle, M. (1996). Clinical gait analysis: A review. *Human Movement Science*, 15(3), 369–387.

Winter, D. (2009). *Biomechanics and motor control of human movement.* Hoboken, NJ: John Wiley & Sons, Inc.

Zelik, K., Collins, S., Adamczyk, P., Segal, A., Klute, G., Morgenroth, D. & Kuo, A. (2011). Systematic variation of prosthetic foot spring affects center-of-mass mechanics and metabolic cost during walking. *IEEE Transactions on Neural Systems and Rehabilitation Engineering, 19*(4), 411–419.

Ziegler-Graham, K., MacKenzie, E., Ephraim, P., Travison, T. & Brookmeyer, R. (2008). Estimating the prevalence of limb loss in the United States: 2005 to 2050. *Archives of Physical Medicine and Rehabilitation, 89*(3), 422–429.

Index

3D 176, 214, 220–222
16 m area 84–85, 91–92, 156–158, 161
30 m area 84–85, 91–92, 156–158, 161

absolute residual 133, 136–137
acceleration 12, 23, 31, 38, 44, 54, 61–62, 64–65, 83, 89, 176, 233, 239–240, 259
actin filament 25–27, 30
action sequence 95–98, 100–105
activation dynamics 37, 43
active phase 53–54, 62–63
adaptation process 187–189, 200
adenosine triphosphate (ATP) 27
advantage of A (AdvA) formula 163–164
aero- and hydrodynamic 6, 13, 56–57
aerobic threshold 175
algorithm 5, 14–15, 58, 73, 78, 87, 92, 110, 112, 130, 154, 176, 188–190, 200
all-out test 202
AMIRA 215, 221
anaerobic threshold 173–175
analysis software 89, 156, 159
anatomy 216, 222
angular motion 62
ANOVA 134, 136
ANSYS Workbench Version 215, 220, 222
anthropometry 214, 216
anticipation of behavior 104
aponeurosis 36
application 14, 16, 18, 22, 40, 51, 55, 59, 63, 74, 90–91, 96, 100, 150, 154, 165, 187, 198, 200–201, 213, 220, 222, 232
artificial intelligence 90
artificial neural network (ANN) 5, 10, 14, 16, 18, 59, 78, 83, 87–88, 90, 95, 97–98, 100–105, 110, 189

assist 108, 111, 113, 115–116, 119, 121, 123, 162, 175
athlete 11, 25, 40, 42–43, 51–54, 58, 60, 66, 95, 99, 147, 149, 152, 169–174, 179–181, 187–189, 196, 199, 201, 212, 237, 255
athlete-boat-oar (ABO) system 57, 60, 62–63, 66
attack 73, 76–77, 81–85, 88, 90, 92, 97–98, 100, 104–105, 110, 147–148, 152, 155–160, 162–163
attacker 100, 104
attribute 18, 52, 78–80, 112, 147–149, 152, 154
attribute vector 78–79, 149, 152–153
axis 32, 56, 238; vertical axis 57

backward fall 212–214, 216–217, 220, 223–224, 230–231
backward selection 59
ball control 78, 80, 91–92, 155–156, 159
ball possession 73–75, 84, 111, 155, 161–164
ball recovery 76, 92, 155
Banister Model 189, 204
Bayesian classification 58
Bayesian nonparametrics (BNP) 173
beach volleyball 104–105
Bernstein's problem 4, 6–7
bicycle ergometer 52, 202
bilateral deficit 43
bilateral extension 43
bio-cybernetic approach 54
biomechanical aspect 60
biomechanical modelling 17, 22, 60, 211–212
biomechanical system 7
biomechanics 10, 51, 53–54, 61, 211, 220, 233

Bionics of Motion 51
black box 3, 15
black-box model 10, 15–16, 18
block 5, 108, 111, 113, 115, 121, 242
boat motion 62–63, 66
body segment 60
Bollay's model 62
bone 3, 211, 213–215, 217–218, 220–224, 227–228, 231–233, 237
brain 5, 7, 13, 90

cadence of paddling 54
calcium 29
calculation 33, 41, 57, 60, 78, 99, 149, 215–216, 220–222, 263
calibration 176, 179, 202
camera perspective 99
campaign 135, 140–141, 143–144
canoeing 50–51
carbon-fiber spring 238
cardiorespiratory system 7
categorical variable 112–114
CATIA V5 214
ceiling perspective 99
center 7, 56, 109, 111–116, 118–120, 123, 150, 213, 245
center of mass (COM) 23, 38–39, 214, 240–241, 245
center of pressure 244, 247, 249, 251
chest strap 176
chi-square value 114
classification 3–4, 10–11, 15, 58–60, 112, 114, 116, 118–120, 152
classification function 112, 114–116
classification model 10, 15
classification system 18
cluster 78, 83, 95, 97–98, 100–102, 111–114, 148–154, 173
cluster analysis 10, 110–114
coach 54, 77, 79, 87, 90–91, 95–96, 109–111, 149–150, 152, 162, 187
coefficient of variation 113
cognitive science 12, 18
competition 11, 25, 40, 50–55, 91, 100, 103, 108–109, 111, 129, 169–173, 175–177, 179, 193, 201
competitor attribute 52
complex (neuro-) musculoskeletal element 60
complex motor task 59
computability of system states 74
computational fluid mechanics (CFD) 57
computational intelligence 58

computational model 6, 12–13, 18
computer aided design (CAD) 220–221
computer aided design (CAD) software 214
Computer Aided Engineering (CAE) 214, 232
computer animation 78
computer calculation time reduction 216
computer experiment 57–58
computer simulation 17, 57, 63
computer tomography (CT) 214, 220
computer vision 99
computer vision system 20, 110
concentric contraction 31
concentric movement 28, 37, 42–43
condition 3, 9–10, 15, 22–24, 31, 33–34, 39–40, 43, 52–55, 57, 63, 100, 143, 149, 152, 162–163, 167, 171, 175, 177, 179–180, 211–214, 216, 220, 223, 227–229, 232–233, 238, 240, 255
constraint 4, 12, 15–16, 138, 184, 192, 195, 199, 255
continuous variable 111, 113, 131
contractile unit 35
controlled space 83
coordinate system 84, 240–241, 243
correlation 42, 75, 89–90, 114, 133, 156, 170–171, 181, 189–190, 199, 201, 240; canonical correlations 114; correlation analysis 89, 159; cross-correlation 189–190; intraclass correlation coefficient 199; Pearson correlation 181, 199
countermovement jump (CMJ) 16–18, 37
court position 103, 109, 111, 113
craft 50
critical area 83–84, 158
cross-bridge 27, 30
cross-bridge theory 25, 34
cross-training 201
cross-sectional area of muscle 24
cross-tabulation 114
cubic model 132, 136
cybernetic modelling 51
cycle ergometer 52, 202
cycle time 54
cycling 201

damping factor 218, 256
data: data accuracy 60; data analysis 113; data-driven modelling 55, 58,

60, 65–66; data pattern 95; data recording 89, 98, 159; data set 59, 100–102, 104, 111–112, 133, 180, 190, 200, 203; data vector 97–98, 103; database 113–114, 172–173, 180, 215; event data 75; dimensionality of the data set 59; empirical data 203; experimental data 65, 218, 223–224, 229; force data 62, 134; Input data 97, 100, 152, 216; output data 97; position data 12, 54, 74, 77–78, 92, 95, 97–101, 103–105, 159
decay/delay factor 198
decision maker 130–131, 144
defense 73, 76–77, 79–82, 104, 108, 149, 155–156
defensive behavior 97, 102
defensive pattern 98, 100, 102–103
deformation curve 249
degrees of freedom (DOF) 3–5, 7, 51, 57, 214, 218
dependent variable 112–3, 132
descriptive discriminant analysis 112
deterministic physical model 10
deviation 36, 65, 89, 101, 113, 136–137, 141, 150, 174, 198–199
dimension analysis 22
dimensionless parameter 23, 41
direct dynamics 24
direction of pass 84
discriminant variable 116
discrimination process 100
distance to goal 100, 102
distance to the nearest defender 100
division 139–144
dorsal splint 219, 225, 227–228, 231–232
dose-response characteristics 190
dose-response model 204
drag coefficient 61–63
drive phase 63
drop jump 234
Dynamically Controlled Network (DyCoN) 11–12
dynamic behavior 56, 83
dynamic model 89
dynamic pattern 74
dynamic process 77, 161
dynamic progression 51
dynamic study 53
dynamic system 56, 74, 77, 90, 188
dynamic training movement 44
dynamical interaction 108

dynamical process 51
dynamics of a system 56
dynamometry 53

eccentric movement 28
effects of training 190–191, 193
efficiency 22, 41–42, 53, 55, 85–86, 92, 155–156, 158, 200, 224
EHF-Men-18-Championship 101–102
elastic property 22, 24, 29, 36, 38–39
elastic structure 22, 28, 36, 42
electrically stimulated muscle 29
electromyography 8, 53
elevated side-view perspective 99
elevation profile 179
empirical internal testing 112
empirical model 23–24, 27
endurance 4, 52–53, 173, 190
endurance athlete 42, 201
endurance sport 197
energy 7–8, 27, 29–30, 42, 169, 176–177, 211, 214, 217, 222, 239–240, 249–253, 256, 258; energy dissipation 255; energy storage/storing 22, 37, 251; kinetic energy 29, 257; potential energy 42, 214, 257
environmental factor 52
enzyme activity 201
equation of motion 23–24, 37, 44–45, 63, 65, 257–258
equipment 3, 23, 54, 170, 209, 211
ergometer 51–52, 202
error 16, 56, 63, 65, 74, 109, 118–120, 174, 180–181; error analysis 216; error rate 113; error tolerance 63; error-based learning 16; least squared error method 174; maximum relative error 56; mean average percentage error (MAPE) 199; measurement error 136, 176, 200; root mean square error 65, 199
Euclidean metric 113
European Committee for Standardization 230
European National League men's handball 96
experimental 55, 130, 223
experimental tournament 130
explanatory model (of movement) 24
"exploitation" 5
external load 108, 222
extreme value 132, 136

factor analysis 10
Fast Fourier Transform 59
fast-twitch fibers 42
fatigue 38, 190–201, 223, 255
feature 3, 10–11, 55, 59–60, 62–63, 65–66, 90, 112–113, 190, 200; feature extraction 59; feature selection methods 59; feature space 112; features characterizing the original data set 59; features controlling 55
feedback mechanism 191
feedback system 180–181
fiber type 22, 35, 42
field goal 108–109, 111, 116, 119, 121, 123
field goal percentage 109
filtering technique 59
finish time 169–170, 180
finite element analysis (FEA) model 215, 220–222, 231
finite element (FE) model 211–212, 222–224, 227, 229, 231
first-order differential equation 192–193
Fitness-Fatigue model (FF Model) 204
foot mechanics 237–238
force: compressive force 255; concentric force 33–34; contraction force 30; dimension of a force 29, 31; driving force 50, 64; eccentric force (f_{ecc}) 33–34; external forces 22, 37, 56, 60, 62, 65; force component 38, 66; force depression 28; force enhancement 27–28, 37; force law 24–25, 37; force plate 240, 242, 245; force vector 240, 243–245, 248–249; force-length relation 28–29, 44; hydrodynamic force 57, 60–62, 65; isometric force (f_{iso}) 23–24, 28, 32–33, 40–42; maximum force 31, 230–232; maximum possible eccentric force ($f_{ecc,max}$) 33–34; peak force 212, 225, 260–261, 263; reaction force (F_t) 23, 239, 258–260; shear force 244; spring force 243; vertical force 17–18, 43, 243, 245–246, 249, 253; vertical ground reaction force (VGRF) 259–260, 263
force-velocity relation 24, 29–30, 32–36, 40–41, 43–44; Eccentric force-velocity relation 33
formation 74, 80–85, 87–88, 90–92, 155, 161, 164
formula of Riegel 170
forward fall 212

forward model 12–14
forward 111–114, 116, 118
foul 108, 111, 113, 115, 121
four-player-chain 76–77
fracture 212, 223–224, 231–232
"freeing" 5
free-throw 108, 111, 115–116, 119, 121, 123
"freezing" 5
friction 29, 31, 41
frog sartorius muscle 41
functional diagnostic 51
futsal 96
Fuzzy set 9

gait 24, 238–240, 255
game analysis 103–104
game analyst 96
game pattern 103
game theory 53, 77, 82, 89, 160
game variable 113–114
game-related statistic 108
game-relevant event 97
genetic algorithm 5, 58
geometric pattern 77, 79, 88, 155
geometrical condition 22
global positioning system (GPS) 54, 62, 176–177; GPS tracking 98
goal time 170–175, 179–181
goalkeeper 100
goal-oriented model 100
graded incremental test 173–174, 179, 181
gravity 22, 57
ground reaction 23, 239, 258–260
guard 109, 112, 114–116, 118–120, 124
gyroscope 62, 242

heart rate (HR) 54, 169–171, 174–184, 201
heart-rate variability (HRV) 201
heat production 29, 42, 211
heel pad 241
heteroscedasticity 132–134, 137
Hidden Markov Models (HMMs) 59–60
hierarchical Dirichlet process (HDP) 173
high-dimensional feature 14
higher-scoring player 112, 114–122
Hill's equation 29–33, 35–37, 40–42, 44
Hill function 189, 194
"hitting the wall" 169–170, 177
holdout 113, 118–124

holistic perspective 121
homeostatic behavior 191
homogeneity 116, 232
homoscedasticity 132–134, 136–137
Hooke's law 24
human (sensori-) motor activity (HMA) 3–6, 10, 18
human interaction 74
human motion 59–60, 211
human motor system 51
human segment model 216
human segment 214
hyaline cartilage plate 221, 229, 233
hydrodynamic drag 57
hyperbolic 32–33, 35–36
hysteresis curve 249

IBM SPSS Software 114
ideal elastic linear spring 24
impact load 217, 231, 238, 255, 258, 259, 263
impedance 239–240, 249, 252, 254
Impulse-Response model (IR-Model) 189, 193
incidence rate 213
inclination angle 38, 40
inclined leg press 38, 45
independent component analysis 59
independent variable 132
indicator 65, 75, 90, 146–147, 156
inertia 37
inertial measurement unit 62
initial activation 45
initial position 23–24, 39, 45, 218
initialization routine 144
injury 108, 212, 214–215, 217–218, 224, 231, 256, 259
input vector 11, 78
inter/intraindividual variability 187–188
interaction of teams 100
internal model 4, 12–13, 15, 179
inverse dynamics 24–25, 37
inverse model 4, 12–14, 16, 136
IR Model 194–195, 199, 201–202
isotonic 29, 31
item process 76
iterative process 78

joint 3–8, 12, 23, 38, 44, 52, 212–213, 217–220, 222, 224–228, 230–234, 237–238, 241, 243–244, 256; joint axes 238; joint center 245; joint load 217; knee joint 9, 241, 243, 256
jump height 43

Kalman filter 189, 195
kayaking 50
key performance indicators (KPI) 75, 146–147, 149–150, 154–156, 158–159, 238, 244, 252, 256; process-oriented KPI 155
kinematics 214, 241
kinetics 193, 241
k-nearest neighbors (k-NN) 59, 112, 114
knee extension 38–39
knockout 20, 129, 131, 138, 141, 144
Kohonen feature map (KFM) 10–11
Kolmogorov Smirnov test 133, 136
kurtosis 133

lactate 169–170, 173–174, 180–181
learning: deep learning 14; error-based learning 16; learning method 8; learning process 7, 152; motor learning 4–8, 13–14, 16, 18; reinforcement learning 14, 16
leave-one-out classification 116; leave-one-out method 112
LED technology 176
length dependence 27–28, 41
length of pass 149, 151
lever arm 246, 248–249, 259
linear actuator 241, 243
linear classification function coefficient 114–116
linear displacement 252
linear elastic elements 36
linear elastic spring 23
linear model 136, 189, 194, 196
linear regression model 65
linear transformation technique 59
load input 178
load-velocity relation 44
"Local Matrix Completion" 172
low-dimensional feature 14
lower-body power 109
lower-dimensional projection 118–119, 121, 123
lower-scoring player 111–112, 114–124

machine-learning 16, 18, 189
macro-and micro-structure 120
magnitude 189, 192, 194–196, 220, 222, 224, 231, 247, 258
manual video analysis 98
manual video tracking 98–99

mass 23–24, 38–39, 44–45, 61–63, 65, 214–217, 230–234, 240–241, 245, 255–258, 260–262
mass of the rower, boat and oar 56, 65
mathematical description 50, 56, 61, 66
mathematical game theory 77, 82
MATLAB 10, 45, 63, 141
matrix 77, 80–82, 152, 154–155, 161, 172
maximal moveable load 29
maximum voluntary effort 24
mean 30, 88, 103, 105, 119, 121, 137, 141, 146, 148–149, 151, 154–155, 170, 172–173, 181, 195–196, 202, 230
measuring performance 110
mechanical model 15–16, 22, 56
mechanical power (p) 31–32, 42, 44
mechanical system 56, 238
mechanical test 237–238, 243–244
mechanical work 29, 42
metabolic system 7–8, 173, 238
methodological experience 15, 22, 25, 60, 74, 87, 98, 113, 132, 159, 162, 175, 181, 191, 243, 256
methods of mechanics 51
metrology 55–56
midfield 74, 90, 162–163
miniaturizing of marker 99
mixed linear modelling 189, 194
mobile device 180
mobile on-board measurement system 63
model of European international soccer 135, 138
model of Keul 174–175
model-fit 190–191, 199–200, 202
modelling technique 58, 129, 132, 134, 188
motion: motion equation 57–58, 62, 66; motion equations' integration 57; motion lab 238–239; motion pattern 16, 55; motion process 59; motion profile 241; motion technique 52–54, 60, 66
motor: motor abilities 3–4; motor approach 4; motor control 3–8, 12, 14, 16, 18, 54; motor program 4–5; motor skill 3–5, 7, 10, 16; motor system 4, 6, 13–14, 16, 51; motor unit 6
moved load 29–30
Movement and Action Sequence Analysis (MASA) 97, 101
movement technique 51, 53–54, 57, 65

multiaxial test machine 243
multibody body simulation (MBS) model 216–217, 220, 224
multinomial 112
multivariate analysis of variance 110
muscle: multipennate muscle 36; muscle activation 37; muscle contraction 22, 26, 30, 217; muscle control 51; muscle cross section 41; muscle efficiency 41–42; muscle fascicle 25; muscle fibers 25, 35–36; muscle force 25, 27–31, 36–39, 41, 43; muscle mechanics 22–23; muscle model 28, 30–31, 33, 36; muscle parameter 22, 25; muscle performance 37; muscle property 22, 25, 38–39; muscle reflex 218; muscle stress 38; muscle-tendon complex 43; unipennate muscle 36
maximal voluntary contraction (MVC) 45
musculoskeletal system 255
myofibril 25, 28
myosin filament 25–27

National Basketball Association (NBA) 109, 111, 113
Navier-Strokes equation 62
navigation 57
nearest neighbor analysis 111–112, 114, 118
network: artificial neural network (ANN) 5, 10, 14, 16, 18, 59, 78, 83, 87–88, 90, 95, 97–98, 100–105, 110, 189; backpropagation network 78; deep network 14; Dynamically Controlled Network (DyCoN) 11–12; feed forward network 78; network process 100; neural network 10, 14, 58, 74, 78–79, 149, 152–155; supervised network 14; training process of the network 97, 101
neural activation process 43
neuromuscular system 22
neuron 5, 11, 59, 78, 83, 97–98, 100–101, 152–154
neurophysiological system 7
Newton's law 23–24, 38, 44, 61–62, 65
Newton-Euler's formalism 61, 211
non-parametric technique 112
normal distribution 112, 132–134, 136–137, 141

oar blade 62–65
occlusion of player 99

offense 73, 75–77, 79, 83, 90, 92, 97–98, 102, 104, 108, 111, 113, 121, 149, 155–156, 159; offense-defense interaction 75; offense formation 80–82, 92, 161; offense type 80, 82; offensive action 83, 97–98, 100, 103; offensive pattern 100–103; offensive tactics 96–97, 102–104
official games 98–99
olympic long jump 43
one repetition maximum (1RM) 44
opposing team 82, 95, 98–99, 160–163
optical sensor 176
optical system 242
optimal length 26–28
optimization 25, 54–55, 57–59, 75, 82, 89, 111, 169–170, 174, 199–200, 203; optimization methods 55; optimization process 199, 203
optional trajectory function 14
orientation 83, 89
oscillation 24, 30, 192, 195, 261–263
outcome variable 131–134
output layer 97–98, 101
overflow 196–197
overflow rate (OR) 178
overload 180, 195, 211, 255
overloading mechanism 177–179
overreaching and overtraining 191

pacing strategy 169–170
paddle 50, 53–54, 57–58, 62, 64–66
parallel damping element (PDE) 36
parallel elastic element (PEE) 36–37
parameter value 23–25, 37, 39–41, 60–63, 65
pass analysis 74, 84, 89
passing path 103
passing position 100
pattern recognition 78, 90, 95, 100, 150
pennation angle 36
percentile, 25th 113
percentile, 50th 215
percentile, 75th 113
performance development 55, 187, 198–199
performance enhancement 37
performance indicators 108, 110–111, 129–130, 237
performance output 178
Performance Potential (meta-)model (PerPot) 170, 177–179, 181, 196–197, 199, 202; PerPot-Run 170–171, 175–176, 180–181
peripheral vascular disease 237
phenomenological model 23
physical curvature 240
physical demand 110
physical feature 52
physical fitness 51, 95
physical model 10, 56–58, 61, 66, 78
physiology 25, 188, 190–191, 195, 211; exercise physiology 188, 191; physiological adaptation 187, 189; physiological phenomena 170, 175, 177, 188
pitch 57, 62, 73, 75, 79, 83–84, 156
plantar flexion 250, 252–253
play analysis 89
player behavior 75, 96; opponents' behavior 53, 155; teams' behavior 89, 147
player-tracking 110
player position 80, 98
PlayGroups subroutine 144
playing court 97
playing pattern 100–102
playing position 112, 114, 116, 118–120
playing process 75, 77, 79, 84, 89, 149, 155–156, 161
playing time 109
PlayMatch subroutine 141
play-off 108, 138, 140, 144
polynomial function 174
position and player combination 96
position information 104
position of the shot 104
position tracking 98
positional-derived variable 110
potentiometer 62
power athlete 42
power output (P) 29, 31
pre-clustering 112
prediction 10, 12–13, 23–24, 65, 77, 104–105, 129, 170–173, 180–181, 189, 200–202, 261; predicted goal difference 137, 141; predicted result 141; prediction analysis 104; predictive equation 112; predictor space 118–119, 121, 123; predictor 118–124, 136
prestart phase 64
principal component analysis (PCA) 10, 59, 110
prior probability of classification 112

probability 30, 35, 53, 55, 60, 90, 95, 102, 104, 112, 116, 130–131, 135, 137, 141, 173; continuous probability distribution 131; discrete probability distribution 131; probability distribution 60, 111, 131–132; probability theory 53; success probability 95, 102
process-based model 73
progression statistic 130, 141, 143
protection 212–215, 224, 231–233
protector 212–214, 218–220, 224–233
prototype 78–79, 83, 153–154
psychological system 7
pylon 239, 243

quadrant map 118–120
quadratic model 132
qualification 129, 144
qualifying tournament 135, 137–140, 143
qualitative model 3
quantification of training load 188, 201
quantitative model 3, 8

random number 131, 135, 141
random variability 134
random variation 130, 134, 137, 141
range of motion (ROM) 220, 231, 253
ranking 73, 129, 139, 143, 147, 149, 152
recovery rate (RR) 178
reference system 56
regatta 52, 54–55, 61–65
regression analysis 10, 129, 133–134, 136
regression model 65, 129–130, 132
relative phase analysis 96
Reserve formula 179
residual sum of squares (RSS) 199–200, 203–204
residual value 130, 132–134, 136–137
resolution 100, 222
respiratory compensation point 175
rhombus 77
Riegel's formula 170–171, 180
rivalry 53, 55
robotic 14, 18
rocker-curve 244
rocker shape 240, 249–250
roll-over 237–240, 244–245, 249–250, 252–253
rotation 57, 213, 218–219, 244, 252–254; center of rotation 242; rotation angle 252; rotational stiffness 252–254

round-robin 192, 131, 138
rowing motion 50, 55, 60
running path 97

sagittal moment 246
sagittal plane 12, 244–245, 247, 249
sampling rate 242
sarcomere 22, 25–28, 35
scalable cluster analysis algorithm 112
scaling factor 192–193, 195–196, 199
scatter plot 133
Schwarz's Bayesian Criterion 111
scoring rate 100, 102
sculling 50, 55, 57–58, 60–61, 63, 65–66
second-order differential equation 23, 61–62
self-organizing maps (SOM) 10–11, 59, 78, 83, 95, 152
semantic similarity 88
sensitivity analysis 25, 40
sensor 5, 7, 12, 59–60, 62–63, 176, 237, 239–240, 242
sensori-motor system 6, 13, 16
sensory feedback 12–13
serial damping element (SDE) 36
serial elastic element (SEE) 36–37
shank 38–9, 60, 66; shank angle 239–240, 242, 245, 252; shank-based coordinate system 240–241
Shapiro Wilk test 133
shear stress 223, 227–229, 232–233
shot 11, 76, 92, 97–98, 100–104, 160, 163; Shot position 100–101, 104–105
shotgun effect 132–133
SIMPACK 215
skeleton 217, 233
skewness 133
sliding filament theory 25, 30
slow-twitch fiber 42
smoothing factor 182–184
soft tissue 211, 217, 222–223, 231, 237, 240
space control 83–85, 88–92, 155–157, 159, 161
spatio-temporal context 89
spectator 91, 96, 144, 169, 171, 180
speed 4, 6, 35, 62, 65, 78, 83, 89, 92, 105, 110, 114–116, 147–148, 155, 160, 162, 169–171, 173–182, 238, 240, 245, 249–250, 258
sport result 52–54
sports biomechanics 51, 54

sports gym 99
spring-damper element 217, 219–220
spring-mass-damper 256–258
squat jump 23
stance 6, 238–241, 249–250, 252, 255
standard deviation 65, 136–137, 141
state of the art 74
static overload 255
statistical analysis 149, 188–190
statistical monitoring code 144
steals 108, 111, 114–116, 123
step width 220
stochastic process 60
strain potential (SP) 178–179
strain rate (SR) 178, 232
strategy 5, 16, 74, 83, 150, 162
strength 4, 15, 52–53, 75, 77, 104, 116, 133–134, 188, 190, 197, 199, 213, 255
stress reduction 228–229, 233
stretch-shortening cycle (SSC) 37, 73
stroke 50, 54, 62–64
structural integrity 256
structural load 261
structural model of the movement 40
sub-clusters 112
subset of variables 112
substantial nonlinear property 110
Support Vector Machines (SVM) 58–59
surface model 214, 216–217
swing phase 237, 241
syntactic and semantic rules 75
syntactic similarity 88

tactic 51–54, 74–76, 80, 92, 95–97, 102, 105, 138, 160, 163; tactical aspects 89; tactical behavior 71, 76, 78, 85, 87–89, 95–96, 98, 155; tactical performance 83, 110, 147
target pattern 103
Taylor expansion 36
team game 79
technical aspect 51, 74, 76, 95
temporal pattern (T-pattern) 95
tendon 22, 24, 36–37
threshold 100, 170–171, 173–175, 180–181, 222
tilt angle 244
time constant 192–193, 197, 204
time frame 199–200
time grid 83, 88
timing 8, 17, 160, 240, 242, 244, 249
titin 22, 26, 29
titin-actin interaction 29

tolerance 13, 15–17, 63, 100
torque 7–8, 23, 52, 217, 230, 239, 244, 252
tracking software 99
training: training concept 74, 147; training effect analysis 187; training effect model 204; training effect 187–192, 199–201; training factor 52, 54; training filming 54; training input 197; training intervention 38; training load 177, 187–188, 191–196, 198–199, 201–204; training process 97, 102–103, 176–177, 190, 197, 199; training vector 78
trajectory 14, 59, 249
transition point 244
trial-and-error 7, 14
trial-and-success 7
triathlon 181, 201
TRIMP 201–202
turnover 108, 111, 113, 115–116, 119, 121, 123
two-dimensional trajectory 59
two-step cluster analysis 111, 113
type A performance model 188, 190–193, 197–202
type B performance model 188, 190–192, 195, 197–201
type of athlete 58
type of motion 58, 66

Union of European Football Association (UEFA) 129–130, 135–136, 138–144
unilateral extension 43
unilateral movement 43
unsupervised neural net 59
upper extremity 60, 66, 212, 214, 217–218

validation 15, 60, 104, 130, 159, 180–181, 201–202, 211, 218, 223–224, 227, 229, 232–233
value sequence 89, 159
Varga's experiment 223–224
Variable Dose-Response model 204
vector 11, 78–79, 97–98, 103, 149, 151–154, 202–203, 222, 240, 243–245, 248–249
velocity: contraction velocity 30–31, 44; corresponding velocity 30; dimension of a velocity 31; initial velocity 12, 23, 39, 258; maximum/largest contraction velocity 31, 34–35; maximum velocity

31, 42; positive velocity 33, 258; shortening velocity 30–31; steady state velocity 30; velocity dependence 22, 29, 33, 36–37, 43
venue 129, 135–137, 141
vertical jump 16, 43
vertical pass 90
video tracking 98–99
viscosity 37, 57
visco-elastic element 218

volleyball 104–105
Voronoi diagram 96
Voronoi cell 74, 83–84, 88, 90, 156–157, 159

white-box model 3, 5, 10, 18
worst performance scenario 113

Z-line 25–27
zone defense 108

Printed in the United States
By Bookmasters